AID & DEVELOPMENT IN SOUTHERN AFRICA:

EVALUATING A PARTICIPATORY LEARNING PROCESS

edited by

Denny Kalyalya, Khethiwe Mhlanga, Ann Seidman, Joseph Semboja

Africa World Press, Inc.

P.O. Box 1892
Trenton, New Jersey 08607
(609) 695-3766

Africa World Press, Inc.
P.O. Box 1892
Trenton, N.J. 08607

First Printing 1988

Typeset by TypeHouse of Pennington, Inc.

Library of Congress Catalog Card Number: 86-73222

ISBN: 0-86543-046-2 Cloth
0-86543-047-0 Paper

Table of Contents

Acknowledgements

This book is the product of the work of several hundred southern Africans in Tanzania, Zambia and Zimbabwe who took part in a year-long regional pilot Learning Process. It explains the reasons for undertaking the process, describes the theoretical underpinning and the country backgrounds, and presents the findings and recommendations.

The dry pages of a book cannot really capture the depth of that process as the participants worked and learned together: the discussions and arguments that dragged into the midnight hours in lamplit rooms or under the trees; the words dropped here, standing with a hand on a plow, or there sitting at a sewing machine; with goodhumored laughter and sometime even tears.[1] We only hope that this book will help them attain their goal of setting up a permanent Learning Process in the region.

It is impossible to express adequate appreciation by name to all the individuals and organizations who helped make the Southern African Pilot Learning Process possible. Oxfam America's Policy Analysis Department, under Larry Simon, initiated the idea of a development assessment strategy and did the spade work needed to set the pilot project in motion, although as soon as possible, the initiative shifted to the southern African participants. The department's advisory board made many useful suggestions in the early stages (see the Appendix for the members' names). Oxfam America staff members and interns[2] spent many hours in the U.S. discussing the basic ideas underlying the pilot process, and helped to gather background information and prepare materials that helped to implement it. In addition, of the Applied Development Research Network,[3] and of Clark University's International Development and Social Change Program, met several times to discuss ways to improve the Process. Some contributed research relative to specific aspects (see footnotes to individual chapters).

A number of private voluntary and non-governmental organizations made valuable contributions to the success of the process, including the Catholic Relief Services; the Canadian University Service Organization; the American Friends Service Committee; the Community Development

Trust Fund of Tanzania; and the Organization of Rural Agricultural Projects of Zimbabwe.

Three individuals took part in the workshops and played an important role in bringing the experiences of grassroots projects in other regions to the southern African Learning Process: Joey Peltier, from Dominica in the Caribbean; and Aruna Roy and Geeta Athreya from India. Their thoughtful comments and suggestions helped the southern African participants realize that they could both learn from and contribute to others' understanding as to the causes and possible strategies for dealing with problems common to many third world countries. Hopefully, this initial contact will lay a foundation for more extended third world linkages to strengthen grassroots projects' collective self-reliance. Most important, we would like to thank the students from Tanzania, Zambia and Zimbabwe who did an excellent job in helping to make the pilot Learning Process work.

Finally, we wish to express sincere gratitude to the foundations and individuals who contributed the funds that made the Learning Process possible. The Nairobi office of the Ford Foundation funded a major share of the southern African costs. In addition, several individual donors contributed generously, including Philippe Villiérs and Richard V. Wolohan and his family.

—The editors:
Denny Kalyalya, Lusaka, Zambia
Khethiwe Mlanga, Harare, Zimbabwe
Ann Seidman, Boston, Massachusetts
Joseph Semboja, Dar es Salaam, Tanzania
1986

Notes

[1]To avoid any possible embarassment, the following text mentions none of the participants by name. The pictures included with the text illustrate the work of typical rural southern African projects, but are not necessarily of people included in the Learning Process.

[2]Particularly Shari Zimble, Portia Adams, Carol Lurie and Dan Bernstein, as well as many members of the Applied Development Research Network and the Clark University International Development and Social Change Program.

[3]See Footnote 3 of Chapter Four.

Foreword

Apartheid in South Africa has been the focus of world attention. South Africa's influence on and disruption of the development of other countries of Southern Africa has been well documented though it has received less attention. This book goes beyond the macropolitical realities to focus on the process of change in small rural communities in Southern Africa.

In 1983, Oxfam American began a study of Southern Africa to look at the impact of long-term development assistance on rural villages. This study breaks away from traditional methodologies to utilize a participatory method, one that involves project holders, and researchers from local African universities. This book sets forth the results of that two-year study.

The book is important for three reasons: it briefly reviews aid policies in Southern Africa; it describes in some detail the participatory methodology called the learning process and it puts forth the results of the study. The latter show that often aid given to rural communities can have unforeseen, and not fully positive, consequences.

It is clear that for development programs to "work" donors must be very sensitive to their role and the role of others in the community. Even when donors want to foster community empowerment, this can be thwarted by the creation of new elites, the exclusion of women, or the antagonism of local government to mention only three. If what donors want is self-reliance and independent community action which the poor control, then donors must be very sensitive and careful to listen to the poor, let go of program control, and be aware of the sociopolitical influence of the development projects.

Over its history, Oxfam America has experimented with a number of research methodologies from totally participatory methods at the community level to the learning process, to participant observation, case studies, etc. Oxfam America has stressed evaluation and research to learn about the process of empowerment. We have found program participants to be our best teachers. By empowering them to talk to us, by listening to people actively, we not only learn for our future work, we build skills to ensure self-reliance.

This is not an Oxfam America publication. The authors are fully responsible for the contents of this book. Yet, the research was undertaken as an Oxfam America program. This book will add to the knowledge of research methodologies, community change in Southern Africa and, most importantly, to the impact of the work of donor agencies in the Third World.

—John C. Hammock, 6/86

Rural southern African women do many kinds of work. (UNICEF photo by Bernard P. Wolff)

Cooperators cultivating their fields. (Photo by Michael Scott, Oxfam America)

Discussing a project proposal. (Photo by Michael Scott, Oxfam America)

Tanzanian rural women in reading and nutrition class. (Photo by Alastair Matheson, UNICEF)

Members of a project planning group. (Photo by Michael Scott, Oxfam America)

CHAPTER ONE
Why Evaluate Aid?

> "What is development? Why is Africa in a crisis after 30 years of aid which supposedly helped the new African states to achieve it? Why has our powerlessness increased? What are we doing wrong?"
> —A participant in the Gwebi workshop, August, 1985.

INTRODUCTION

This book reports on a pilot attempt to involve southern African aid recipients in a participatory Learning Process to evaluate the impact of aid on their efforts to attain self-reliant development.

Over a quarter of a century after the first sub-Saharan colony won its freedom, some fifty new African countries have attained political independence. Through bilateral and multilateral government channels and private voluntary organizations, Western countries multipled the dollar amounts of aid. But by the 1980s, the high hopes born of the end of colonial rule had mired in confusion and despair. Not only in Southern Africa, but throughout the entire African continent, people found themselves engulfed in crisis.

As an introduction, this first chapter

- Summarizes some data relating to African aid, debt and living standards;
- Outlines the history of recent trends and the debates concerning bilateral aid provided by the United States, as well as multilateral aid through the World Bank and the IMF;
- Briefly describes the growing emphasis on private voluntary organizations in the aid process, noting that they, too, have not been entirely free of criticism;
- Discusses the growing concern with the need for evaluation to enable aid givers and aid recipients, alike, to learn from their experience in order to improve the impact of aid.

AID AND DEBT

The Southern African Learning Process created only a small window through which to look at the impact of private voluntary aid on a few projects in one region. Nevertheless, it may give an insight into the factors influencing the consequences of the much larger amounts of aid provided to Africa by the international community.

Table 1.1 shows the growth in official bilateral and multilateral aid as defined by the World Bank.[1]

TABLE 1.1. Total Outstanding Bilateral and Multilateral Official Aid[a] to Sub-Saharan Africa, 1970–1982 (in US $ millions)

	1970	1975	1980	1981	1982
Concessional bilateral loans	2,377	5,319	10,173	10,906	11,566
Official export credits	324	927	5,505	5,606	6,187
Multilateral loans	842	2,529	8,443	9,752	11,414
Privately guaranteed loans	1,876	5,330	16,521	18,185	18,895
Grants (net disbursements)	n.a.	3,169[b]	7,149	7,067	7,183

Notes:
[a]The World Bank defines official development assistance (ODA) as net disbursements of loans and grants made at concessional financial terms by official agencies of the members of the Development Assistance Committee (DAC), of the Organization for Economic Cooperation and Development (OECD), and members of the Organization of Petroleum Exporting Countries (OPEC), with the objective of promoting economic development and welfare. It includes the value of technical cooperation and assistance.
[b]1976.

Source: World Bank, Toward Sustained Development in Sub-Saharan Africa - a joint program of action, (Washington, D.C.: World Bank, 1984)

In the 1980s, as the economic crisis deepened, African states joined other Third World countries in long queues seeking loans, accumulating further debts and deepening their dependence on foreign assistance. Although in absolute terms, the African states obtained less than other developing countries (as Table 1.2 shows), their debt burden weighed more heavily on their economies. By 1984, it totaled two-thirds of their aggregate Gross Domestic Product, compared to little over one-third for all developing countries. Putting it another way, African countries' foreign debt exceeded two times the value of their exports compared to only one and a half times for all developing countries. To pay off their accumulated debt, African countries would need to expand exports faster than the other countries.

Like their sister states further north, as they attained independence, the southern African states obtained increased aid from the developed world. By the late 1970s, despite its declared determination to achieve self-reliance, Tanzania received one of the continent's highest per capita

TABLE 1.2. Africa's External Debt Compared to That of All Developing Countries (in US $ billions, as a percent of Gross Domestic Product, and as % GDP and of exports[a])

Region	1977	1980	1984
Africa (excluding South Africa):			
External debt (US $ billions)	112.2	87.4	169.5
As % of GDP	35.8	35.7	61.8
As % of exports	135.8	143.2	215.8
Debt service payments (US $ billions)	11.0	13.3	26.3
All developing countries:			
External debt (US $ billions)	333.5	566.8	829.5
As % of GDP	24.7	24.9	37.9
As % of exports	125.0	109.0	145.4
Debt service payments (US $ billions)	39.4	87.7	128.6

Note: [a]Includes goods and services.

Source: World Economic Outlook - A Survey by the Staff of the International Monetary Fund, 1984.

amounts of aid. Zambia, with a relatively high income per head that concealed pervasive rural poverty, found it necessary to borrow heavily abroad simply to cover its current expenditures. Within a year after attaining independence, Zimbabwe invited potential donors to its widely-publicized Zimcord Conference; it sought some $2 billion in aid to offset the damage caused by the liberation war and the decades of poverty imposed on its population by minority rule. (See Table 1.3)

In the 1980s, the nine southern African states[2] joined together in the Southern African Development Coordination Conference (SADCC) to accelerate their pace of development and reduce their dependence on South Africa. Yet even that collaborative endeavor depended heavily on foreign aid. SADCC planners, for example, anticipated that foreign sources would supply 87 percent of the finances required for their proposed food security program, and all the funds they needed for agricultural research.

Despite expanded exports and the increased flow of aid, the majority of Africans experienced declining incomes and worsened living conditions. (See Table 1.4) Formerly self-sufficient at least in agricultural produce, the continent began to import foodstuffs. Nevertheless, tens of millions of children, women and men barely survived, chronically on the verge of starvation. Foreign aid helped provide "modern" education and medical techniques, but every year tens of thousands of malnourished children died before the age of five.

TABLE 1.3. Foreign Debt and Debt Service Costs of Independent Southern African States, 1978 and 1982[a]

	Foreign Debt (in US $ millions)		Debt Service Cost (in US $ millions)		Debt Service as % of export earnings	
	1978	1982	1978	1982	1978	1982
Angola	378.0	2,262.0[b]	n.a.	361	n.a.	20.5[c]
Botswana	120.7	202.0	n.a.	n.a.	1.9	2.0
Lesotho	32.7	138.6	3	26	38.2	58.9
Malawi	n.a.	692.0	n.a.	n.a.	8.6	23.0
Swaziland	105.2	177.7	n.a.	n.a.	1.7	3.6[b]
Tanzania	1,142.0	1,632.0	38	113	6.1	n.a.
Zambia	1,449.0	3,294.0	194	642[d]	21.7	67.0
Zimbabwe	418.0	1,221.0	8	146	0.8	13.9

Notes:
[a]Refers only to external public debt; inclusion of private external debt—for which data are unavailable—would significantly increase the debt service ratio.
[b]1981.
[c]Using 1981 export figure, and 1982 debt-service payment.
[d]Repayments due in 1983.

Source: Economist Intelligence Unit, Quarterly Economic Reports for Each Country, 1983 (data not available for Mozambique); 1982 debt service ratio for Zambia and Lesotho calculated from the International Financial Statistics (Washington, D.C.: International Monetary Fund, December, 1984)

THE DEBATE

The crisis of the 1980s heightened the ongoing debate over aid and its impact on development. Much of that debate, focusing on official aid, reflected an underlying theoretical disagreement as to the cause of poverty and underdevelopment and the kinds of strategies required to provide jobs and rising living standards for Third World poor.

Mainstream theories that undergird most official aid from the United States and Western Europe tend to view Third World underdevelopment as the result of lack of capital, skills, entrepreneurship and markets. Aid, they conclude, must fill those gaps.[3]

Critics of aid, however, offer three sets of interrelated explanations for aid's failure to contribute to development. First, some argue that much official aid stems not from a desire to help the poor, but from a desire to shape Third World development to meet the developed countries' interests. A 1984 article in a Zimbabwe magazine revealed an increasingly common cynicism:

> Aid is not charity. . . .

TABLE 1.4. Some Economic Indicators of the Worsening Conditions in Africa, 1967–1985

	1967-76	1980	1981	1982	1983	1984	1985
Terms of Trade (% change)[a]	3.1	15.2	2.9	−4.9	−3.2	0.6	−2.3
Balance of Payments (as % of exports)[b]	−15.8	−4.9	−27.5	−31.4	−20.6	−15.1	−10.3
Net investment income[c] (US $ billions)	n.a.	−4.6	−5.2	−6.5	−6.7	−7.0	−8.2
Debt service ratio	n.a.	13.3	15.1	19.4	22.9	26.3	32.4

	1960-70	1970-80	1981	1982	1983
Per capita income:					
% average growth, sub-Saharan Africa	1.3	0.7	−0.9	−1.7	−2.0

Notes:
[a]Changes, in percent, compared to previous year.
[b]Balance of payments on current account as percentage of total exports of goods and services; minus shows deficit.
[c]All investment income, except payments of income on foreign direct investment; includes receipts from direct investment abroad. Minus shows outflow.

Sources: International Monetary Fund, World Economic Outlook, 1985 (Washington, D.C.: IMF, 1985) is the source of all data except for per capita income, which is from World Bank. Towards Sustained Development in Sub-Saharan Africa, op. cit.

> In the newspapers and the mass media in general, as well as in learned papers written by intellectuals, aid is almost always presented as if it will feed the drought-stricken poor of the Third World, and build their roads, power stations and agricultural farms. This presentation of aid may make it look good in the eyes of both the recipients and the donors, but in fact, it is an upside-down presentation of reality.
>
> The reality is that the primary motivating force that moves aid originates not from the recipient but from the donor countries. Officially, bureaucratic practice dictates that the initial request for aid must originate from the recipient country. But before an official request is made, there is always prior discussion . . . (C)oncretely the choice of programmes and projects, which a donor country takes up for consideration, are those which they hope would advance their interests.[4]

Those adopting this view maintain that many kinds of strings are attached to official aid. As a price of their aid, governments may insist on "reforms" which—failing to take into account local conditions or historical reality—may undermine Third World countries' capacities to solve their own social, political and economic problems. Some critics claim the donors may impose conditions designed to chain recipient nations to their political, economic, and even military objectives.

The second set of explanations for aid's inability to foster development relates to the donors' failure to involve recipients—especially the rural poor—in planning and implementing aid projects. Those holding this view argue that much official aid consists of centrally-designed, large-scale development projects which exceed small countries' institutional capacity to absorb them. These projects rely on sophisticated technologies requiring the import of machinery, equipment, and highly paid expatriate managers. They tend to increase rather than reduce the local population's external dependence.

A third set of explanations focuses not on the donors' aid policies, but on recipient countries' misuse of aid.

To provide some evidence relating to the first two explanations, the following section outlines the changing trends in the United States' bilateral aid, and the multilateral programs of the International Monetary Fund (IMF) and the World Bank. It touches only briefly on the issues relating to Third World countries' misuse of aid.

Official United States aid

Although as a percent of the Gross National Product its aid has been declining, the United States remains the world's largest official donor (see Table 1.5). As such, it plays a major role in shaping official aid's impact on Third World economies. A review of its aid policies in Africa may provide a perspective on the first two sets of explanations of official aid's impact on development.

Prior to World War II, the US had few economic, political or strategic interests, or historical ties in Africa. As the first African states attained independence in the late 1950s, however, the US rapidly expanded its aid to the region. The US normally accompanied the establishment of diplomatic relations with new governments with bilateral aid programs designed to build infrastructure and foster growth. Initially, these sought to ensure stability of governments sympathetic to US perspectives, and, over the longer term, to facilitate US firms' access to African markets and resources. In emergencies, the US also provided humanitarian relief.[5]

As the Vietnam War absorbed more attention and resources in the mid-1960s, US aid to Africa declined. Only in the late 1970s did the US restore its economic assistance to Africa to the 1963 level. The Sahel drought of the 1970s spurred a modest increase. At the same time, a congressional "New Directions Mandate" targeted US AID priorities to the poor majority, emphasizing participation as a major theme: US AID should actively engage the poor, including women, in decision-making and implementation in ways which increase "their technical skills and/or their capacity to organize for common purposes and for greater access to the benefits of development."[6]

Available evidence, however, revealed gaps between these new concepts of participatory planning and the realities of implementation.[7] US AID remained accountable only to Congress and agencies like the Office of Management and Budget, not to poor villagers.

In 1981, the US administration emphasized two aspects of its aid to Africa. As one observer noted,

> These . . . reflect the ideological predilections of the Reagan administration—an emphasis on the East-West conflict and US strategic and political interests worldwide *vis-a-vis* the Soviet Union, a skepticism regarding the effectiveness of aid as a tool of development, and a belief in the efficacy of free markets and the private sector in stimulating economic expansion.[8]

First, the US administration shifted its aid focus from "equity" and concern with the basic needs of the poor, toward growth, promoted by private sector investments and economic policy "reforms." US AID established a Bureau of Private Enterprise. It failed, however, to convince

TABLE 1.5. Official Development Assistance Provided by the United States Compared to Other Major Donors, 1960-1981 (in US $ millions and as a percent of the national Gross Domestic Product)

	1960		1970		1980		1981	
	$m	%/GDP	$m	%/GDP	$m	%/GDP	$m	%/GDP
United States	2,702	.53	3,153	.32	7,138	.27	5,760	.20
United Kingdom	407	.56	500	.41	1,851	.35	2,194	.43
Japan	105	.24	458	.23	3,353	.32	3,170	.28
France	823	1.35	971	.66	4,162	.64	4,022	.78
Netherlands	35	.31	196	.61	1,630	1.03	1,510	1.08
West Germany	223	.31	599	.32	3,567	.43	3,182	.46

Source: World Bank, The IDA in Retrospect, Oxford, 1982

US private enterprise to expand investments in Africa; in fact, total foreign investment there declined as crises spread throughout the continent.[9]

Second, the US administration emphasized the use of aid to promote US security and political interests. In southern Africa, this took the form of "Constructive Engagement," aimed at persuading the minority ruling South Africa to introduce compromise reforms to end apartheid, while leaving intact the basic socio-economic structures.[10] In the Horn of Africa, it meant construction of US bases and increased military support for Sudan, Somalia and Kenya. From 1980 to 1985, US military assistance multiplied 150 percent—far more than the 40 percent rise in bilateral economic assistance which failed even to keep pace with inflation. As a result, military assistance rose from 10 to 20 percent of total US bilateral assistance to the region.

The US AID director to Zimbabwe drew on the Zimbabwe experience to illustrate the effect of the official US political stance on aid.[11] By the end of 1984, Zimbabwe had received $293 million from the United States, about 10 percent in the form of loans. Part of the grants financed imports from the United States, for which private Zimbabwe firms paid in local currency; the Zimbabwe government, with US agreement, could spend these funds on local development projects.[12] The US AID director explained that the United States sought to help create a pluralistic and democratic society in Zimbabwe as a model in the southern African region. He pointed out, however, that

> The difficulty . . . is that if the initial or basic rationale that goes behind the aid is political, then the aid programme can become a political hot potato and that's exactly what happened a year ago.

According to this AID director, the US representative at the United Nations, Jeanne Kirkpatrick, kept a voting record of all the countries around the world:

> (H)er score card was not very favorable to Zimbabwe . . . (and the situation) was made worse when Zimbabwe abstained from the vote condemning the Soviet Union for shooting down the Korean airline (sic-eds.). And then there was the invasion of Grenada. Most countries either voted against the US or abstained but Zimbabwe with Nicaragua actually sponsored the resolution condemning the US, and it was that step that caused the cut in aid.

The Reagan administration shaped other features of bilateral aid to conform with its ideological predilections. In 1985, the US administration proposed a new Economic Policy Initiative to spur Third World nations to adopt more market-oriented policies. Some critics observed that the limited funds—$75 million in the first year, and $100 million in the next four years—could exert little influence for making comprehensive reform. Instead, the administration would more likely use them as contingency funds to advance its political/strategic goals. Others argued that, even if the initiative accomplished its stated goals, the thrust of the proposed reforms—centered on reduced African state participation in the economy—ignored the reality shaped by a century of colonial rule: In Africa, only the state has the capacity to redirect investments to more balanced, integrated development capable of fostering increased employment and higher living standards.[13]

US Economic Support Funds, at $391 million in 1985, remained the largest element of bilateral aid. Their flexibility and speed of distribution made these funds desirable for all African states in urgent need of finance for imports and budget support. But the administration concentrated almost 60 percent of them in the countries of the Horn of Africa—Sudan, Kenya and Somalia—where the US has important strategic interests.

The Development Assistance Program, usually directed toward projects that take four to eight years for completion, received less funding in 1985 than 1984. Dispersed more widely throughout the continent, it provided an average of $10 million a year to 30 sub-Saharan African countries. In contrast, the administration allocated about 20 percent to Kenya, Sudan, and Somalia.

Food for Peace constituted another major vehicle for US emergency relief. Only pressure from the public and Congress, however, pushed up the 1985 levels of food aid to avert mass starvation in Ethiopia.[14] Prior to that, the administration had allocated the largest amount of food aid ($50 million a year) to Sudan, with programs of between $16 and $20 million to Somalia, Liberia and Zaire.

Not only did the US shift the focus of its bilateral aid, it also sharply reduced its contribution to multilateral agencies. A memo from the Office of Management and Budget Director to the President proposed a halt to all US contributions to multilateral institutions. This raised so many objections from developed and developing countries alike, that the administration set the proposal aside. Nevertheless, the US reduced its contribution to the World Bank's "soft window," the International Development Association (IDA).[15] This resulted in the reduction of IDA loans to African states—which had received about 40 percent of its credit—from $1.6 billion to $1.2 billion. The US administration also pressured the World Bank and International Monetary Fund (IMF)—in which it holds almost a fourth of the votes—to impose market-oriented conditions on recipients in return for their assistance.

Multilateral aid agencies—The World Bank and the IMF

As the largest contributor to multilateral aid agencies, the US government may exert a significant influence on their aid policies. The World Bank and the International Monetary Fund (IMF) constitute important multilateral sources of financial assistance for African as well as other Third World countries. The IMF lends member countries funds to offset international payments deficits. In the 1980s, as African countries' exports declined and inflation raised the costs of their imports, they, like other Third World countries, relied more heavily on the IMF for assistance. (See Table 1.6)

When a member country requests funds from the IMF in excess of its quota, the IMF may require it to make certain "reforms" as a condition for further assistance. Drawing on monetarist theories, the IMF staff hold certain conditions essential to enable a country to balance its international payments and pay off its creditors. The "reforms" typically include: 1) increased government austerity involving cuts in government employment, eliminating subsidies designed to keep food prices down, and reducing social services; 2) reducing government intervention in the economy; 3) freezing wages throughout the country; 4) creating an attractive climate for foreign investors by reducing their taxes and permitting them to remit more profits home; 5) raising the bank rate to reduce inflationary pressures presumed to arise from an expanded money supply; and 6) devaluing the national currency to reduce imports and expand exports.

Many economists and government officials have criticized IMF "conditionality" on the ground that its "reforms" rest on abstract theoretical models which fail to take into account the real causes of the problems confronting Third World countries.[17] In many, if not most, cases where governments have complied, the people have mobilized widespread demonstrations against the resulting price increases, growing unemployment and worsening social services. If a government rejects the IMF's conditions, the IMF may refuse assistance. Since most international commercial banks follow the IMF's lead, the government then faces acute difficulties in obtaining funds to finance essential imports: the oil, machinery and equipment, even foodstuffs, on which its "modern" sector depends.

Unlike the IMF, the World Bank provides longer-term loans. As Table 1.7 shows, these have constituted an important source of funds for African countries' development programs.

The World Bank Charter, however, allows Bank loans only to projects

TABLE 1.6. International Monetary Fund (IMF) Credit to Africa especially Tanzania, Zambia and Zimbabwe, Compared to the Developed Countries and the Rest of the Third World, 1947-1985 (In US $millions and as percent of the total[a])

Country	1947-79	1980	1984	Total 1947-84	% of all countries
Africa[b]	3,780.5	874.1	1,225.9	12,496.0	13.6
Tanzania	138.6	40.0	—	202.3	0.2
Zambia	439.0	50.0	147.5	1,225.7	1.3
Zimbabwe	—	32.5	89.9	313.4	0.3
All developing countries	20,672.0	3,752.7	8,076.5	62,068.3	67.4
Industrial countries	29,282.0	—	28.5	29,739.5	32.3
Total	50,210.2	3,752.7	7,081.7	92,065.9	100.0

Notes:
[a]The principal way the IMF makes resources available to countries is to sell currencies of other members or SDRs (Special Drawing Rights) to them in exchange for their own currencies. A member country is said to make "purchases" or "drawings" from the Fund. Once a member has spent its quota of its own currency in the funds, it must obtain credit from the fund to purchase additional currencies of other countries. Since the Fund's resources are of a revolving character, used to finance temporary balance of payments deficits, members must subsequently repurchase their currencies from the Fund with the currencies of other members or SDRs. The IMF reports monthly on each member country's purchases.
[b]In 1982, the IMF permitted South Africa to purchase an equivalent of SDR 902.2 million, equal to 34 percent of all currencies allocated to Africa, and raising the African share of the worldwide purchases from the IMF.

Source: International Monetary Fund, International Financial Statistics, 1985.

TABLE 1.7. Loans to Sub-Saharan Africa by the World Bank and the International Development Association, 1970-1982 (in US $ millions and as percent of total official public and publicly-guaranteed loans)

	1970	1975	1980	1981	1982
World Bank	590	1,261	2,549	2,854	3,327
IDA	226	880	2,573	3,090	3,728
Total	816	2,141	5,122	5,944	7,055
As % of total official credit	15.0%	6.9%	12.6%	13.3%	14.6%

Source: Calculated from World Bank, Toward Sustained Development in Sub-Saharan Africa, *op. cit.*

which supplement and do not compete with private sector investments.[18] In Africa, the World Bank at first primarily funded infrastructure for large-scale foreign mining investments. In the 1970s, it shifted more funds toward export agriculture, even financing small peasant farmers' efforts to expand cultivation of export crops. Like US AID, the World Bank began to stress greater peasant participation in designing and carrying out the programs it funded. Its 1978 Annual Report observed:

> . . . with hindsight, project design and the pace of implementation have been too ambitious, resulting in delays and shortfalls from original expectation . . . (A)mong the more difficult aspects is the establishment of systems within which small farmers can themselves have a say in how programs are designed and implemented, and how their skills, expert knowledge of the local farming environment, and their capacity to help themselves can be fully integrated into an overall effort.

An unpublished 1977 review of 164 World Bank rural development project appraisal reports concluded that the majority contained only the most minimal data on the social, demographic and economic characteristics of the project area. They made no attempt to analyze the implications of such data for project design. The review added that Bank staff had done little to utilize the potential of indigenous social organizations.[19] Even the Bank's Mexican PIDER Project, committed to incorporating extensive popular participation in decision-making and implementation, turned out to involve "little more than wishful thinking, since no local mechanisms had been developed to give reality to the ideal."[20]

By the 1980s, as African countries imported more and more staple foods for their growing urban and export-oriented sectors, the World Bank called for greater emphasis on food cultivation. At the same time, however, the United States reduced the resources available to its "soft window" affiliate, the International Development Agency (IDA), and exerted growing pressure for privatization of production activities. In the 1980s[21] the World Bank began to press for structural adjustments. These encompassed decreased state intervention in domestic markets and foreign trade, and greater support for foreign as well as domestic private investments.[22]

Is foreign investment aid?

Most African economists objected that pressures by multilateral aid agencies to privatize the typical African economy would likely aggravate, rather than alleviate, the causes of underdevelopment.[23] They rejected the notion of foreign investment as "aid." They pointed out that all African states as they achieved independence sought to attract industry to provide employment, tools and equipment to manufacture low-cost consumer necessities and increase productivity in all sectors, including agriculture. Transnational corporate decision-makers, however, tended to invest, if at all, in the more profitable, relatively more developed sectors of the African economies: the "modern" enclaves geared to crude mineral and agricultural exports; and import-substitution industries that used capital-intensive imported machinery to process imported materials for narrow high-income groups.

As an extreme reflection of this tendency, US manufacturing firms located three-fourths of their investments on the entire continent in South African factories[24] under the then-profitable conditions of apartheid. In the rest of Africa, foreign investors failed to invest in employment creation and rural development.

Critics of foreign investment argued that foreign investors drained more surplus out of Africa than they contributed in the form of new capital. For example, in almost every year from the late 1950s (when the first sub-Saharan African states won independence) to 1985, US firms *extracted* from African countries more profits, interest and dividends than they *invested.*[25] Moreover, by the late 1970s, private foreign investments in Africa had declined absolutely.[26]

In sum, the evidence relating to official aid from the United States and multilateral agencies reveals contradictory trends. On the one hand, they began to call for increased participation by Third World countries in the design and implementation of rural programs. On the other hand, especially in the 1980s, they began to press African states to make structural adjustments to open their economies to private investment and "market forces." Critics objected that private investment tended to render African political economies more, rather than less, vulnerable to the impact of international crises.

Exploitation of aid by local elites?

The third set of explanations for aid's failure to foster development in Africa relates to its alleged misuse by government officials and private profiteers. Zimbabwe's US AID Director claimed that "most of the resources have been squandered or misused, in most cases misused to support a minority elite."[27]

Although such a sweeping generalization appears overstated, evidence exists of misuse of aid by individuals in both the private and the public sectors of some African countries.[28] Once a steady flow of aid becomes institutionalized in a national economy, it may generate its own elite among those who distribute aid within the country.[29]

Critics from the left[30] and the right[31] allege that state marketing agencies sometimes set unrealistically low prices for peasant food crops for one of two reasons: to skim off the surplus to support the bureaucracy; or to keep down urban food prices to retain the allegiance of civil servants who comprise the bulk of the more vocal urban workers. Low urban food prices may discourage peasants from growing more food crops. Therefore, the governments may use food aid, not to feed starving rural inhabitants, but to stock urban markets with food, the sale of which provides an additional source of revenue.

In short—although this is by no means always the case—elites may misuse aid to line their own pockets or enhance their own power and prestige.

PRIVATE VOLUNTARY ORGANIZATIONS AS AN ALTERNATIVE

Mounting criticism of official aid has contributed to growing pressures for channeling aid through private voluntary organizations (PVOs). Proponents of this approach argue that private agencies have greater potential than large government bureaucracies for empowering people at the grassroots to achieve self-reliant development. Private agencies are less likely than large, impersonal official donor agencies to "throw money" at problems; they are more likely to encourage local populations to participate in program formulation, enabling them to improve their own skills and to learn to employ their own resources. These arguments, among others, helped to persuade the U.S. administration and Congress to direct US AID to channel an increased share of total US aid funds through private voluntary organizations.

TABLE 1.8. Private Voluntary Organization Grants as Percent of Total and United States Official Development Assistance (ODA) and Other Official Flows (OOF) (bilateral and multilateral)

	1970	1975	1980
Total PVO grants (in US $ millions)	857	1,342	2,371
U.S. PVO grants (in US $ millions)	598	804	1,301
Total PVO grants as % of ODA & OOF	9.7	7.5	6.9
U.S. PVO grants as % of U.S. ODA & OOF	15.7	14.0	13.6

Source: Adapted from the Organization for Economic Cooperation and Development's "Cooperation Efforts and Policies of the Members of the Development Assistance Committee," Reviews, 1976-1981.

Private voluntary organizations development aid efforts, however, have not escaped criticism. Frances Korten suggests that, regardless of a declared preference for participation, persisting bureaucratic obstacles inherent in the centralized service-delivery approach of donor agencies may thwart involvement of rural peoples in program design and implementation. This is because the centralized, service-delivery approach:

(i) locates decision-making relating to budgets, equipment and personnel at the center. Official rhetoric or even genuine policy decisions can have little effect if the agency leaves this decision-making structure intact.

(ii) assumes the donor has technical and even general knowledge to impart to improve rural peoples' lives. Underlying this assumption is the implicit belief that the agency, not the people, "know" what the people need; therefore the staff need not waste time consulting them.

(iii) provides a multitude of signals to its employees about what it expects of them: data to collect about their work, the targets set, the training provided, the status accorded various positions, the avenues for promotion, pay scales, and the content of supervisory discussions. These determine staff members' trade offs in deciding how to go about their work—and typically leave little time for encouraging rural peoples' participation.

(iv) expects personnel to carry out standard activities, regardless of location, so individuals can be transferred from one post to another with relative ease. This contradicts the need for continually deepening the staff's acquaintance with rural inhabitants' circumstances and institutions in order to effectively engage them in improving their skills and using their own resources to solve their problems.

Frances Korten emphasizes that a large agency bureaucracy may find it difficult to change these characteristics of the centralized, service-delivery approach; but that unless it does, no matter how much its staff may publicly proclaim their desire, it will prove incapable of involving rural people in a participatory approach.[32]

David Korten notes, "(T)here is little evidence to suggest that, when undertaken on anything approaching the scale required, private voluntary efforts are consistently more effective than those of government."[33]

Others argue that aid to individual projects, of the type typically provided by private voluntary organizations, cannot, by itself, lead to self-reliant development for the larger community. Unless the project receiving aid fits into local and national plans, it may fail to achieve anything more than helping a few individuals learn to sew, or plant and sell crops. It may even foster emergence of a new elite. Even if the project becomes integrated into a larger community effort, poorly designed and implemented national and international programs may exert a negative impact which leaves the community worse off than before.[34]

THE NEED FOR EVALUATION

The growing criticism of aid, whether channelled through official agencies or private voluntary organizations, has stimulated recognition of the need for systematic evaluation and learning from past mistakes. Most official government aid agencies, like US AID, and multilateral aid agencies, like the World Bank, have instituted various forms of reviewing their projects and programs. These typically require periodic reports from staff members. Frequently, they employ outside evaluators.

Private voluntary organizations, too, generally require their staff to report on the progress of projects. They have completed financial audits and monitoring, but not in-depth evaluation.

In a review of US and Canadian nonprofit organizations, Brian Smith concludes that they have failed to develop adequate evaluation techniques for several reasons:

> 1) they have generally not devoted sufficient time or resources to conduct a thorough evaluation; 2) they employ inappropriate standardized techniques for small-scale projects; 3) they have not devised methods capable of assessing intangible results like enhancement of hope and self-esteem or political and social awareness; 4) evaluation by outside personnel can undermine the trust between donors and recipients; and 5) they tend to rely on evidence from the organization's staff and the projects' leaders, rather than non-leader beneficiaries or community members outside the projects.

Smith observes that the unevenness and inadequacy of their evaluations do not help private voluntary organizations to prove their claims that they are more effective than government agencies in providing more innovative techniques, fullfilling basic needs, involving recipients in decision-making, or enhancing the bargaining power of low-income people. Equally, if not more important, they have failed to institutionalize a systematized internal learning process or feedback loop from past experiences to present and future decision-making.[35]

By the 1980s, African research personnel and members of grassroots projects throughout the continent had begun to object that too often foreign "experts" visited the projects for only a few days, talking mostly with the leaders and looking mainly at physical aspects. Often the "experts" could not speak the local languages and lacked basic information about the communities' cultures. Written primarily for donor organizations, the reports often viewed issues from the agencies' rather than project members' perspectives. Although the reports' contents frequently influenced the donors' aid policies, project members seldom saw them. Project members learned little from the evaluators about how to improve their own activities, and far less about the possibility of using local resources. Qualified African researchers, excluded entirely from these evaluation exercises, could neither learn from nor contribute to gathering information and designing more effective development strategies.

Early in 1984, staff members in planning and research for some 30 US private voluntary organizations met in Washington, D.C., to discuss the need to improve evaluation techniques.[36] All expressed concern that their existing evaluation methodologies failed to provide adequate feedback as to the effectiveness of their aid. Many emphasized the worsening national and international conditions confronting the projects to which their organizations transferred resources. Some scored official bilateral and multilateral aid programs which impacted negatively on rural peoples. Several objected to policies like "Constructive Engagement" in Southern Africa, the growing militarization of aid in areas like the Caribbean Basin and the Horn of Africa, and the austerity conditions imposed by international financial institutions.

The Washington meeting concluded with a call for private voluntary organizations to work together to devise new innovative evaluation methodologies to assess the consequences of aid for grassroots efforts to achieve self-reliance. Oxfam America reported on the proposed Southern African Pilot Learning Process Project which aimed to provide a participatory approach for evaluating the consequences of private voluntary organizations' aid for grassroots development.

The Learning Process concept is not new. Its roots reach back into the notion of learning-by-doing emphasized by such widely different theoretician-practitioners as John Dewey, Paolo Friere, Karl Popper and Karl Marx. Some years ago, David C. Korten[37] observed that "social development does not lend itself to the conventional developing programming methods which call for experts to design program blueprints to be passed to line agencies for implementation." Instead, he asserts, it requires "a bottom-up learning process by which the program design and the capacity to implement it are developed simultaneously to produce a three-way fit between the beneficiaries, the program, and the assisting organization . . ." Korten formulated a model to illustrate the way the elements of this kind of Learning Process should meet in a three way fit.[38]

As Korten defines the Learning Process, however, it differs from the one adopted by the Southern African Pilot Learning Process. Korten placed his emphasis on the goal of attaining a "fit" between the three elements. In contrast, the Southern African Pilot Project focused on institutionalizing an on-going participatory evaluation process through which the project members, as well as the donors, could learn from the mistakes they would (inevitably) make in implementing their program. Thus it aimed to create the necessary framework within which the project members, step-by-step, could enhance their self-reliant capacity to assess and improve their own skills and resources to fit their needs.

The rest of this book summarizes the basic theory, findings and conclusions of the Pilot Southern African Learning Process Project. Chapter 2 outlines the theoretical foundations of the problem-solving methodology employed to involve project members, together with national researchers and donor agencies, in evaluating the impact of the aid they receive from several private voluntary organizations. Chapter 3 analyzes the socio-economic background of the three countries—Tanzania, Zambia and Zimbabwe—where the pilot Learning Process took place. Chapter 4 describes how the participants designed and implemented the process.

Chapter 5 reports on the main findings. Chapter 6 summarizes the participants' conclusions as to how to improve and institutionalize an on-going Southern African Learning Process to strengthen the project members' capacity to evaluate aid.

CHAPTER ONE
Notes

[1]The measure of the aid transfered to Africa depends, of course, on what its definition includes. Most people include all grants of funds and the cost-free transfer of technical skills. In reality, however, this constitutes a relatively small share of the aid available to African and other Third World countries. Most authorities include loans, although some emphasize that the higher the interest charged, the less the aid component. Some define private investments as aid. A few include military assistance.

[2]Angola, Botswana, Lesotho, Malawi, Mozambique, Swaziland, Tanzania, Zambia and Zimbabwe.

[3]For example, the World Bank reported at the beginning of the 1980s that, for low-income African countries to achieve middle-income investment levels, they would require foreign resources to finance as much as 40 percent of their total investment (World Bank, World Development Report, 1980. Washington, D.C.: International Bank for Reconstruction and Development, 1980).

[4]"The hidden face of aid," MOTO, Nov. 1984, p. 4.

[5]Unless otherwise cited, for this brief review of US aid policies in Africa, and the subsequent shift to a growing emphasis on political and strategic concerns, see Carol Lancaster, "US Aid, Diplomacy and African Development," Africa Report, July-August, 1984.

[6]Agency for International Development, Implementation of "New Directions" Development Assistance, Report to the Committee on International Relations on Implementation of Legislative Reforms in the Foreign Assistance Act of 1973 (Washington, D.C.: U.S. Government Printing Office, 1975) pp. 7-8.

[7]*cf.* David C. Korten, "Community Organization and Rural Development," Public Administration Review, Sept/Oct, 1980, p. 483. The New Directions mandate followed criticisms of the earlier focus on economic growth which failed to spread benefits to the poor. See also Irma Adelman and Cynthia Taft Morris, Economic Growth and Social Equity in Developing Countries (Stanford, CA: Stanford University Press, 1973); Montague Yudelman, "The World Bank and Rural Development" in G. Hunter, A.H. Bunting and A. Boltrall, eds., Policy and Practice in Rural Development (London: Overseas Development Institute, 1976).

[8]Lancaster, *op. cit.,* p. 63.

[9]World Bank, Accelerated Development in Sub-Saharan Africa (Washington, D.C.: World Bank, 1981) Table 3.1, p. 17.

[10]See Ann Seidman, "The Roots of Crisis in Southern Africa" (Africa World Press, for Oxfam America, 1985) for discussion of Constructive Engagement and its impact on southern Africa.

[11]"US AID's aid programme in Zimbabwe," MOTO, Nov. 1984.

[12]Although in the 1970s Kissinger had talked of $2 billion to help solve the problem of white settlers who owned the country's best farm lands, the Zimbabwe government could spend none of these funds for land resettlement. (See New York Times, Sept. 9, 1976; also "Let's Build Zimbabwe Together: ZIMCORD" Conference Documentation, Zimbabwe Conference on Reconstruction and Development, Salisbury, March 23-27, 1981, p. 1).

[13]This does not imply that the African governments have always pursued appropriate policies, but only that—given the historically-imposed constraints—private investments alone cannot achieve the essential reconstruction of the inherited economy. For discussion of this reality, see Ann Seidman, *Planning for Development in SubSaharan Africa* (New York: Praeger Publishing Company and Dar es Salaam: Tanzania Publishing House, 1974).

[14]Prior to the 1974 ousting of Emperor Haile Selassie and the imposition of a major land reform, the United States had allocated a major share of its aid to Africa—including military assistance—to Ethiopia. After the revolution, it shifted the focus of its assistance to neighboring Somalia, Sudan and Kenya.

[15]In the mid-1960s, the World Bank set up IDA to provide loans at little or no interest charge to countries considered among the world's poorest.

[16]Lancaster, *op. cit.,* p. 64-5.

[17]*e.g.* Charlotte Peyer, *The Debt Trap and the IMF* (New York: Monthly Review Press, 1975); see also Graham R. Bird and Tony Killick, *Quest for Economic Stabilization: The IMF and the Third World* (New York: St. Martin's Press, 1984).

[18]For a useful review of World Bank policies, including the inherent conflict between the criteria mandated by its Charter and the Bank's focus on participation, see R.L. Ayres, *Banking on the Poor—the World Bank and World Poverty* (Cambridge, Mass.: MIT Press, 1984).

[19]See Robert S. Saunders, "Social Analysis in Rural Development Projects: A Review of Bank Experience," and "Traditional Cooperation, Indigenous Peasant's Groups and Rural Development: A Look at Possibilities and Experiences," August 29, 1977, unpublished, cited by D. Korten, *op. cit.* p. 483.

[20]Michael M. Cernea, Measuring Project Impact: Monitoring and Evaluation in the PIDER Rural Development Project—Mexico, World Bank Staff Working Paper No. 332 (Washington, D.C.: The World Bank, 1979) cited in D. Korten, op. cit., p. 483.

[21]A.W. Clausen, the former head of the world's second largest commercial bank, the Bank of America, assumed the Bank presidency in 1981.

[22]World Bank, Annual Reports, 1980.

[23]Cf. UNU-UNITAR-TWF Reports, "Future Strategies of Africa" Project, Dakar, Senegal, 1980-1984.

[24]United States Department of Commerce, Current Survey of Business (Washington, D.C.: Government Printing Office, August, 1979).

[25]U.S. Department of Commerce, Survey of Current business, *op. cit.,* annual reports on foreign investment.

[26]World Bank, Accelerated Development in Sub-Saharan Africa, *op. cit.* Table 3.1, p. 17.

[27]in *Moto, op. cit.,* p. 9.

[28]In some countries, including Zimbabwe government personnel allegedly have sold food aid for their private profit. See Andrew Meldrum, "Food Relief Fraud in Zimbabwe," Guardian (England), May 19, 1984.

[29]In Somalia, for example, the employment and incomes of a significant number of bureaucrats reportedly depends on the continued flow of aid. ("Somalia Keeps Its Clutches on Aid for Ogadan Refugees," *Christian Science Monitor*, Dec. 8, 1981)

[30]*eg.* Samir Amin's TWF-UNITAR-UNU papers on "The Peasants and the State."

[31]*eg.* The World Bank, Accelerating Development in Sub-Saharan Africa, *op. cit.*

[32]Frances Korten, "Community Participation—A management perspective on obstacles and options," in David C. Korten and Felipe B. Alfonso, eds., *Bureaucracy and the Poor* (New York: McGraw-Hill International Book Co., 1981) pp. 183-189.

[33]David Korten, "Community Organization and Rural Development," *op. cit.,* p. 438.

[34]This latter point aroused considerable discussion at the final regional workshop of the pilot project (See Chapter six below). National intermediary spokespersons from as widely separated locations as Zimbabwe, India, and Dominica emphasized the need to relate individual projects to overall community development.

[35]Brian H. Smith, U.S. and Canadian Nonprofit Organizations (PVO's) as Transnational Organizations (New Haven: Yale University Institution for Social and Policy Studies, PONPO Working Paper, 70, and ISPS Working Paper, 2070. 1983).

[36]The meeting, held at the InterAmerican Agency's Washington, D.C. headquarters, was arranged by staff members concerned with evaluation from several organizations, including PACT and Oxfam America.

[37]See "Social Development—Putting the people first" in Korten and Alfonso, eds., *op. cit.*

[38]See *ibid.,* pp. 213ff.

CHAPTER TWO

Toward A Participatory Evaluation Methodology

> You only get really self-reliant development when the people themselves raise questions and examine the causes of the difficulties they face and look for new answers. You can't push projects down their throats.
>
> —A participant in the Gwebi Learning Process workshop, August, 1985.

This chapter:

- Outlines the aim of the pilot Learning Process; and
- Explains its three theoretical foundations: participation by project members; a problem-solving methodology; and use of national researchers.

THE AIM OF THE PILOT LEARNING PROCESS

The pilot Learning Process aimed to involve the project members and aid agency staff in assessing their projects. It did not focus on another critical aspect of aid, that of the decision-making process within the international private voluntary organizations or the intermediary agencies through which they may transfer their aid. The process contributed nothing to analyzing how the staff of international organizations or their intermediary implementing agencies decide which projects to assist, and what kinds of resources to transfer to them. To clarify the limited aim of the pilot process, Figure 2.1 pictures the several actors in the aid process.

International private voluntary organizations, intermediary agencies, and projects, all have their own perspectives and goals. These may not always coincide. They also have their own internal decision-making structures:

- International private voluntary organizations, like other bureaucracies, may respond, not only to the concerns of Third World rural inhabitants, but to the pressures of donors and internal structures shaped for centralized service delivery.[1]

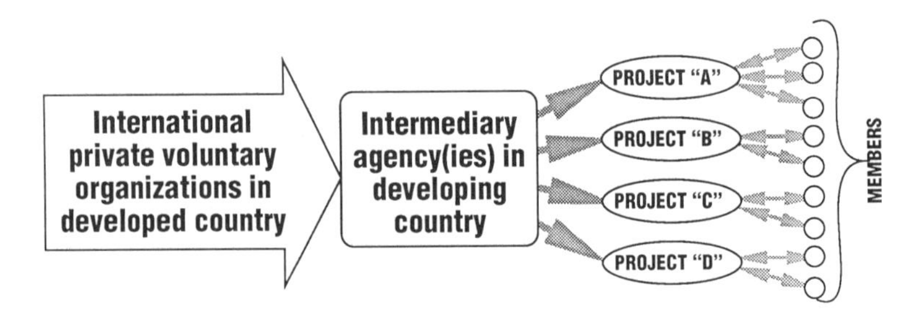

FIGURE 2.1: The several actors in the transfer of aid.

Graphics by William Z. Harris

- International private voluntary organizations, like other bureaucracies, may respond, not only to the concerns of Third World rural inhabitants, but to the pressures of donors and internal structures shaped for centralized service delivery.[1]
- International private voluntary organizations frequently transfer resources through a variety of intermediary organizations in the differing socio-economic contexts of particular Third World countries. These intermediaries may include the international organizations' field offices, or the field offices of other international agencies with similar perspectives. Local personnel, expatriates, or both may run these offices. International organizations not infrequently transfer aid through non-government groups of nationals (sometimes with religious affiliations, sometimes secular) who seek to stimulate rural development. Occasionally, they work through local development agencies with fairly close ties to government, as long as these share similar development perspectives. Each of these intermediaries has its own priorities and methods of work.
- The projects receiving aid may each have their own goals which implicitly, if not explicitly, may differ from those of the intermediaries and donors. Their individual members, too, may have conflicting needs and demands which the projects' internal decision-making structures may or may not adequately mediate.[2]

The pilot Learning Process focused on how projects affected the way members utilized aid in their efforts to achieve the projects' stated goals. Figure 2.2 adapts a decision-making model to illustrate the Learning Process function in providing feedback to the project members and aid agency staff. The model depicts the way the decision-makers behave in a particular institution or conversion process, using inputs, conversion and feedback processes to produce a given range of outputs. Continuous feedback should provide information influencing the decision-makers' behavior, suggesting how they may improve it to achieve the desired goals.

The model depicts three sets of activities to produce outputs: i) input processes, that is, the transfer of aid along with other resources to the project; ii) conversion processes, that is, the way project holders work together to utilize the resources; iii) feedback process. If the outputs do not correspond to the goals stated in the initial project document—and they almost never do—the feedback process should help to explain why and lay the basis for decisions leading to improved future performance.[3]

The model focuses on the way project members respond to their range of choices in using the resources transferred to them by the aid agency. Although it provides no insight into the factors determining the donor agencies' decision-making processes, improved information as to how the project members are likely to behave in response to the new opportunities created by the transfer of resources should enable them to devise aid strategies which better meet the project members' needs.[4]

By systematically examining how project members function to attain stated goals, the Learning Process also aimed to relate theory to practice. To illustrate: Most private voluntary organizations generally hold that their aid empowers project members to attain self-reliant development. Admittedly, "self-reliance," both as concept and theory remains poorly defined.

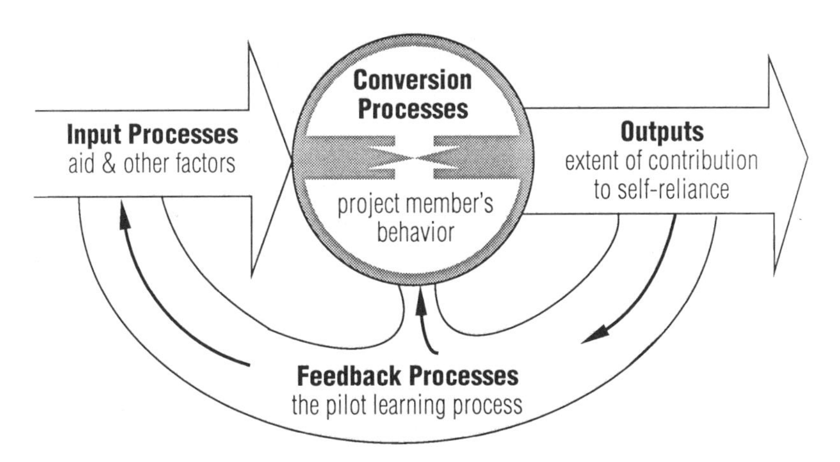

FIGURE 2.2: A model of the role of the learning process in the transfer of aid.

Graphics by William Z. Harris

Systematic evaluation of projects that receive aid, however, should help the project members and aid agency staff to identify both the patterns of behavior (institutions) and resource allocations that affect attainment of self-reliance. By increasing their understanding of the constraints and resources, the process should empower the project members to deal more effectively with their turbulent environment.

At the same time, the process should help further define the concept of self-reliance and develop the theory explaining it. Just as the collapse of a bridge raises questions not only about its plan and construction, but also about the underlying theory of mechanics on which its design rested; so, research which evaluates aid also should test and help to improve not only specific plans, but also the underlying theory on which the aid is based.

Thus, the pilot Learning Process sought to utilize an approach to evaluation which would contribute to empowering project-holders to attain self-reliant development. Three primary features constitute the theoretical foundations of the learning process: participation by the project holders together with donor and intermediary agencies; a problem-solving methodology; and the role of qualified facilitators, preferably nationals.

THE THEORETICAL FOUNDATIONS

1. The participation of project members: ROCCIPI

In recent years, more and more evaluators have emphasized the advantages of involving those whose work is being assessed in the evaluation process.[5] The Learning Process aimed at testing the proposition that through participation in an improved evaluation process, project holders can learn to analyze and overcome the obstacles thwarting their efforts to attain self-reliance.

A considerable body of theory undergirds this proposition.[6] The southern African pilot project sought to adapt that theory to the process of evaluating aid projects. To plan aid, donors must, of course, anticipate how, within the constraints and resources of their environment, project holders will behave.

The argument for project members' participation in the Learning Process rests on two premises. First, common sense argues that no one can better describe those constraints than the project members themselves. By creating an environment in which the members participate in the evaluation process, the voluntary agencies can tap their special knowledge of the causes of their problems.

Second, by engaging in a systematic evaluation of the obstacles hindering their progress, the project members may acquire a better understanding of their own behavior as well as of the constraints and resources within which they hope to attain their objectives. In the process, they should learn to formulate more self-reliant strategies for the future. Examination of the seven categories of factors likely to influence project members' behavior in their efforts to attain project goals supports this proposition.[7] These categories may be remembered by using the mnemonic, ROCCIPI: Rule, Opportunity, Capacity, Communication, Interest, Process, Ideology.

RULE: Like any law or norm, the project document typically prescribes the changed behavior required to achieve the stated goal: The members must work together in certain specified ways to sew uniforms, to plant, harvest and sell crops, etc.

OPPORTUNITY: By transferring specified resources, aid aims to create the opportunity for the members to use it, together with other resources, to achieve their goals.

CAPACITY: Unless they have the necessary skills, however, members will likely miss their goals. In addition to the ability to sew, or knowledge about the ways to plant, harvest and sell crops, they must have (or the transfer of resources must help them acquire) bookkeeping and management skills. They must also have internal institutions which enable them to make appropriate decisions and settle disputes. Finally, external factors must not thwart the possibility of their successful achievement.

COMMUNICATION: For the project members to behave in appropriate ways, they must have received and understood information about the project goals and processes. To ensure this, communications theory underscores the advantages of engaging the members in a two-way, fact-to-face dialogue.

INTEREST: Members probably will not behave in appropriate ways unless convinced that the project serves their interests.[8]

PROCESS: Members will more probably change their behavior to carry out the project as planned if they have participated in the decision-making process. Since evaluation constitutes a crucial feature of the decision-making process, project holders should participate in evaluation as well as the initial planning process.

IDEOLOGY: If project goals or process contradict the community's values and attitudes (what Alvin Gouldner terms "domain assumptions") project members will probably not behave in appropriate ways to achieve the goals, unless through participation in an evaluation process they consciously decide to adopt different values and attitudes.

These seven categories provide an agenda or checklist for evaluation research designed to explain the project members' behavior in utilizing aid. The factors they encompass all lie within the close environment of the project members, who therefore possess special knowledge concerning them. For these reasons, the Learning Process focused on creating the opportunity for grassroots project members to participate in analyzing them. By involving the project members in the design and implementation of the process, it aimed to help them to understand and view evaluation as essential to their own interests.

Furthermore, the final category, ideology, suggests that project members' own values may impose a major constraint on their development efforts; unless they change them, they may behave in ways that impede attainment of their stated goals. For example, many southern Africans, accepting the traditional belief in women's inferiority, exclude them from project decision-making. However, women grow 40 to 80 percent of the food crops, depending on where they live in the region. Unless women understand and participate in formulating plans for projects to grow food crops, those projects may fail.[9] If, by participating in the Learning Process, the project members themselves discover that their traditional exclusion of women from decision-making blocks their progress, they will more likely seek ways to include them.

In sum, the pilot Learning Process aimed at enabling project members to strengthen their capacity to plan and improve their own efforts to achieve self-reliance. At the same time, the process sought to create a mechanism for channelling their findings to the private voluntary agency community to enable it to improve its contribution to their efforts.[10]

This analysis may help to explain not only the disillusionment with outsiders' evaluations, but also why many projects fail to attain their goals. First, without the members' participation, evaluators can only with difficulty discover the causes of the problems encountered. Second, unless they participate, the members, themselves, may never understand the causes. Without that knowledge, self-sustaining development becomes an unattainable ideal.

As Mamdani put it:

> (A)ny strategy that claims to be a solution must seek to revive the creativity and initiative of the people. Central to this must be educating people about the relations that make them disaster-prone. This education must be based on investigation, concrete and independent. And it must lead to organization, both popular and around concrete issues.[11]

2. A problem-solving methodology

To achieve meaningful project member participation in the evaluation process requires an appropriate methodology that fosters learning through doing. For this purpose, the Southern African Pilot Project adopted a problem-solving approach.

The problem-solving methodology differs qualitatively from the ends-means approach employed by many evaluators to assess aid projects.[12] The ends-means approach typically takes the goals of the project as given. Frequently, this implies stating the goals in quantitative terms: In a given period, for example, the plowing of so many acres and the production and sale of so many bags of maize, cotton, beans or other crops; the sewing and sale of a stated number of school uniforms; the digging of a specified number of boreholes. The evaluators then assess the means which led to success or failure as measured by the extent to which the project achieved those quantitatively formulated goals.[13]

The ends-means approach stems from the widely-used but seldom explicitly stated theoretical framework which pervades conventional decision-making.[14] By its very nature, the ends-means approach cannot serve voluntary agencies' requirements. It identifies ends with values, about which people may debate, but which publicly available data cannot serve to validate. At the end of the day, this approach tends to take values and even institutional structures as reflecting "society's" goals. As a corollary, ends-means proponents argue that, except by surveying opinions, evaluators cannot even conduct empirical research concerning values and therefore goals. This frequently implies that an aid project can only encompass incremental change. It must leave intact the basic institutional behaviors and attitudes of society, although these are often interwoven into and sustain the fabric of underdevelopment.[15]

As a logical consequence of its underlying premises, the ends-means approach limits decision-making to the mobilization of bias: Those with the power to make decisions—whether on a national, a village or a project level—may impose their values on the less powerful. For example, with the best will in the world, a private voluntary organization may simply adopt conventional wisdom as to what project members should do to achieve self-reliant development. Since it has the power to give or withhold aid, it may thus shape the project's goals.

At the level of the project, members may accept the leaders' decisions because, for whatever reasons, the community institutions give them the power. But the members who disagree may implement the decisions made only woodenly. As ROCCIPI's agenda suggests—and as the swath of failed aid projects in the development field attests—an ends-means approach that neglects the role of member's values and institutions cannot fulfill the aid givers' dreams.

It is possible to distinguish three types of feedback: goal seeking or negative; goal-changing or learning; and consciousness and self-awareness feedback.[16] We can understand all three by analogizing to a navigator's use of a compass. In the first type, the compass tells the navigator she is off course, so she shifts her rudder to correct; this negative feedback simply tells the decision-maker what action to take to return to the initial course. In the second type, the navigator looks up from her compass to see that, following her charted course, she will collide with an iceberg; so she charts a new course (changing her goal) to avoid it.

These two types of feedback characterize the likely consequences of the ends-means approach to evaluation for the aid process. The decision-makers have selected the goal; they seek feedback only to learn if the course they have charted is being followed or, more rarely, whether new circumstances require them to change the goal.

The third type of feedback probes to reveal less obvious obstacles which may nevertheless pose serious dangers. For example, the navigator may discover that local magnetic factors have affected the needle of her compass, so she must obtain a new gyro-compass to avoid being thrown onto the rocks. Decision-makers need this third type of feedback to expose the way existing institutions themselves (not merely poor decision-making within those structures) block attainment of planned goals. This type of feedback develops "consciousness" or "awareness," since it may reveal the necessity of changing the institutions governing the project members' responses.

To obtain this third type of feedback, the pilot Learning Process adopted a problem-solving methodology.[17]

Unlike the ends-means approach, the problem-solving methodology provides a framework for engaging the decision-makers in systematically analyzing the causes of the difficulties they confront. It seeks to identify the resources on which they may rely, and the obstacles—including institutions and values—which may constrain their efforts. Thus it lays a basis on which they may identify strategies to overcome the institutional and resource constraints which might otherwise obstruct their goals.

Adapting the problem-solving methodology to evaluating the impact of aid requires a recognition that aid, itself, constitutes a proposed partial solution to the problem of underdevelopment as perceived by its designers—whether these be the aid agency, the project members, or all of them together. For example, food-for-development aid aims to overcome the problem of malnutrition and, in the process, spur food production. A credit project seeks to overcome the difficulty posed by the peasants' inability to raise capital to buy oxen, tractors, fencing, or improved seeds and fertilizers. A small rural industry manufacturing hoes aims to provide rural industrial employment, as well as end peasants' dependence on imported equipment that requires scarce foreign exchange.

Like all human enterprise, development—whether on the national or a grassroots level—is an on-going process. No planned strategy ever succeeds perfectly. Both internal factors and changing external circumstances inevitably impede its progress. Underdevelopment persists because of these circumstances. To conclude that projects have failed because they encounter difficulties in achieving their initially-stated goals reflects an inadequate understanding of the development process. Instead, the project members, together with those who assist them, must continually monitor the results of implementing aid strategies. In this process, they will inevitably discover new difficulties. They must then analyze the causes of these new difficulties as the basis for revising their old strategies or formulating new ones. Gradually, thus, the project holders themselves will improve their capacity to use their own resources to achieve self-reliant development. To put it another way, they will learn, in the process, how to surmount the difficulty which led to the initial project.

The pilot Learning Process, therefore, sought to adapt the five basic steps that comprise the problem-solving methodology to its proposed participatory process for evaluating aid. It aimed to engage the project holders together with donor and intermediary agency representatives to:

1. IDENTIFY THE PROBLEM: Define the nature and scope of the difficulties or problems that hinder project members' efforts to attain their goal. The Learning Process involved the project holders themselves, since, as the ones most affected, they know the most about the problems they confront.

2. CONSIDER THE FULL RANGE OF EXPLANATIONS: Formulate all possible explanations of the problems into consistent sets of propositions capable of being tested in light of objective information available from the project. All too often donor agencies and project holders assume they "know" what causes the difficulties. As in the ends-means approach, their conventional wisdom tends to take society's values and institutions as given. In contrast, systematic consideration of all possible explanations will more likely discover unsuspected casual factors. The attempt to formulate the candidate explanations as a set of logically consistent propositions capable of being tested against

the facts, facilitates determination as to which ones deserve further analysis. Involvement of the project members in this exercise will enable them to learn how in the future to analyze better the causes of their problems.

3. TEST THE EXPLANATIONS AGAINST THE FACTS: Involve the project members in gathering evidence to test which of the alternative possible explanations coincides most closely with the evidence as to the causes of the difficulties the project confronts. Social science cannot "prove" the truth of a proposition, but it can determine which candidate explanation seems most consistent with the available evidence. The formulation of explanations as testable propositions simultaneously suggests which relevant facts project holders should gather to test them. By emphasizing project holder understanding as critical to project success, ROCCIPI underscores the necessity of engaging project holders in gathering relevant information to evaluate the validity of the alternative explanations. In the process, they will acquire more knowledge of the resources available to them, as well as the constraints likely to thwart their efforts.

4. PROPOSE SOLUTIONS: Discovery of the explanation most consistent with the available data helps to empower project members to devise better strategies for solving their problems. It exposes the causes of the difficulties which they must address. Having participated in the first three steps, the members should have acquired the new understanding needed to devise more suitable strategies to overcome those causes.

5. MONITOR THE IMPLEMENTATION OF THE NEW STRATEGY: Since the new or revised strategy will inevitably encounter further difficulties, the problem-solving approach emphasizes the necessity of institutionalizing an on-going participatory feedback mechanism. Having taken part in the first phase, the members will have acquired the skills for this exercise. In other words, participation in the problem-solving evaluation process will help to empower them to conduct an on-going evaluation of their next steps, a vital foundation for building self reliance.

In short, the problem-solving methodology provides an agenda for integrating on-going evaluation into the design of every grassroots project, each of which constitutes a part of the overall complex, contradictory process of development. The problem-solving methodology provides all those engaged in transferring aid to grassroots projects with an opportunity to understand the factors hindering its role in empowering the rural population to achieve self-reliant development; and gives them concrete tools for improving its impact. In this sense, the methodology provides a systematic framework for involving all the actors in the transfer of aid in an on-going "learning process."[18]

3. The use of national researchers

Conducting a participatory evaluation of aid projects does not mean simply bringing the project holders together in a workshop, explaining the process and leaving them to implement it. That approach would assume that bringing people together in a kind of quilting bee to discuss their mutual problems would create a group dynamic, enabling the participants to work together to find more effective solutions to their problems. It would ignore the need for a scientific methodology and knowledge of the full range of possible causes that may block formulation of more effective development strategies.

An experiment with self-evaluation in Zimbabwe illustrated the difficulties of simply encouraging project members to undertake a self-evaluation. Prior to the introduction of the pilot Learning Process, one intermediary agency sought to encourage the members of the same group of projects to conduct a self-evaluation. A staff member met with the leaders of the projects, explained the purpose of the exercise, and left them with a questionnaire to fill out with the information they gathered. Examination of the "answers" provided revealed that few, if any, of the project leaders had really understood the aims of the exercise, far less learned anything from it. They merely listed some incomplete facts about the background of the participants in the projects.[19]

A brief review of ROCCIPI helps to explain why, without on-going participation by a facilitator skilled in a problem-solving methodology, a participatory evaluation will likely fail:

> RULE: The pilot project proposed that those engaged generate statements of difficulties, explanations and proposals for solution;
>
> OPPORTUNITY: The Learning Process provided the opportunity for implementing the project;
>
> CAPACITY: The problem-solving methodology requires the ability to formulate and test the full range of possible explanations of the causes of rural poverty and underdevelopment which aid aims to empower the project members to overcome. The causes of problems affecting projects may exist at several levels: i) the lack of management capacity,

with all the technical skills that aid may help to provide; ii) factors hindering the democratic participation of all the project members in project decision-making; and iii) externally imposed constraints, ranging from government policies to the consequences of the international recession. An understanding of the multiple causes of underdevelopment calls for a fairly high level of social science education denied to most rural dwellers around the world.

COMMUNICATION: A workshop involving the project members would not suffice to communicate to them all the necessary understandings and skills required to produce meaningful explanations and proposals for solution; that would require an on-going, learning-by-doing kind of process throughout a prolonged period facilitated by someone capable of assisting them to "tease" the relevant explanations out of their complex environment.

INTEREST: Some donor and intermediate agency staff members, not to mention some project leaders, might not find it in their interest to explore all the relevant explanations. It might turn out that their own negative role causes the difficulties. To leave the evaluation process to those who may seek to utilize aid for their own ends in these circumstances would likely prove counterproductive.[20]

PROCESS: Under the best of circumstances, bringing donor staff and project members together in a participatory process is difficult. Each has unconscious attitudes and patterns of behavior that may constitute blocks to effective participation. Without an outside facilitator, these may remain as obstacles to implementation of a truly participatory process.

IDEOLOGY: The attitudes and values of the project holders themselves, as well as donor and intermediary staff members, may thwart needed behavior changes to implement the proposed learning process. For example, if project holders' traditional attitudes exclude women from participation in decision-making concerning a food crop project, that same bounded rationality may hinder them from recognizing their exclusion as a major cause of failure. A sympathetic outside facilitator might help them to discover and perhaps overcome this constraint on their efforts.

Added together, these suggest compelling reasons for including a facilitator to assist project members to develop a participatory learning process.

Widespread criticism of aid agencies' employment of expatriate evaluators constitutes an important factor leading to the design of the pilot Learning Process. In part, this criticism may reflect as much those evaluators' non-participatory style of evaluation as the fact that they were expatriates. In part, however, it reflects a two-fold concern: First, expatriate evaluators frequently lacked sufficient familiarity with the culture of the region. Second, exclusion of national researchers from the evaluation process denied them the opportunity to learn from, as well as contribute to, grassroots rural development. As the Nigerian Ambassador to the United States observed, if aid agencies do not include African researchers in finding solutions to underdevelopment, they may become part of the problem.[21]

In the early years of independence, because of colonial neglect of education, donor agencies could argue with some justification that not many Africans had acquired the needed skills to conduct the kinds of evaluation they sought. However, by the 1980s, over a quarter of a century after the first African states had achieved independence and begun to build up their own educational institutions (including universities), that argument no longer had validity. Africans constituted a high proportion of the social science staff in most southern African universities.[22]

Given that the development of a participatory evaluation process, at least at the outset, requires a perspective and facilitation from outside the project, engaging African nationals as facilitators, rather than bringing in foreign "experts," has significant potential benefits. First, African researchers more likely speak the project members' language and have greater knowledge of their cultures and traditions. They can work closely in a participatory process with the members, helping them to strengthen their capacity to analyze and find solutions to their problems.[23]

Second, as staff members of national research or teaching institutions, national facilitators can integrate the results of the information gathered in the course of the evaluation process into the expanding national body of critical thought needed to ensure more effective planning, not only at the grassroots but also at the national and even the regional levels. Over time, through participation by their staff in the Learning Process, national research and teaching institutions may obtain valuable grassroots evidence to test the broad range of development theories. This should contribute to improving these theories as possible guides for the formulation and implementation of national plan strategies, helping to reduce national as well as small rural projects' dependency on outside aid.

For example, peasant cultivation and sale of export crops like cotton may fail to generate anticipated increased cash incomes because of an oversupply of fibers (including synthetics) on world markets, or a world recession may reduce their prices. Recognition of this reality may spur the project members to cultivate crops less dependent on the vagaries of external markets. However, expanded markets do make possible the specialization and exchange necessary to increase productivity and raise living standards. National research and teaching institutions may draw on the grassroots projects' experience to design more effective plans for specialization and exchange in the context of more balanced development.[24]

In other words, the participation of national researchers as members of national research and teaching institutions in developing an on-going participatory learning process should strengthen the national (as well as the project members') capacity to analyze the causes of poverty and underdevelopment and plan more effective strategies to overcome them.[25] Given the fairly recent origin of southern African institutions of higher education and research, this proposition may be more true there than in Asia or Latin America. First, in Africa, many of the national researchers come from rural backgrounds that enable them to empathize with and understand peasants' problems. Second, the relative scarcity of high-level personnel has limited the division between teaching and research that characterizes institutions of higher learning in some other regions. This means that researchers will more likely introduce the lessons of the learning process into classes they teach. They may also integrate the learning process methodology into teaching curricula, helping to reduce, rather than widen, the gap between theory and practice, between the lives of rural people and the teaching and research community.

The pilot Learning Process involved an effort to work closely with the national research and teaching institutions to select as facilitators researchers who not only possessed relevant academic qualifications, but who empathized with and understood the villagers and their problems.

SUMMARY

The Southern African Pilot Learning Process Project aimed to design a participatory, problem-solving methodology. It created a framework within which members of rural projects, together with national researchers and donor agency representatives, could evaluate the consequences of aid for the project members' efforts to understand and improve their use of available resources to achieve sustained on-going improvements in their incomes and living conditions.

Viewed this way, the Learning Process constitutes more than an insightful "window" on the project for the donor agency. It contributes to the creation of new decision-making structures which empower project members to deal more effectively with their environment. That, after all, is what aid is all about.

CHAPTER TWO
Notes

[1]Frances Korten discusses the obstacles to donor agency support for participatory project development in "Community Participation: a management perspective on obstacles and options" in David C. Korten and Felipe B. Alfonso, eds, *Bureaucracy and the Poor: Closing the Gap* (New York: McGraw-Hill International Book Co, 1981); the arguments are summarized in Chapter 1 above.

[2]Frances Korten examines several obstacles which may hinder participatory organization of community structures. *ibid.*

[3]M. Baratz and P. Bachrach, in *Power and Poverty* (New York: Oxford University Press, 1970) criticized Dahl's model as static. This model attempts to introduce a dynamic character by emphasizing decision-making activities or processes.

[4]The model may be easily adapted to consider the role of feedback in aid agencies by substituting them (and their internal staff structures) for the project members in the conversion process: the input and feedback processes define the factors likely to influence their decisions; and the outputs constitute their transfer of particular resources to projects.

[5]*E.g.,* see David C. Korten, "Community Organization and Rural Development: A Learning Process Approach, in Public Administration Review, 40, 1980, pp. 480-503 for review of some case studies illustrating the advantages of increased participation in project design and evaluation; also see comments in Chapter 1.

[6]See Korten, "Community Organization and Rural Development," *op. cit.* An international participatory research network has been established, with participants from Africa, Asia and Latin America, which has been experimenting and developing the theory and practice of participatory research. Its head office is in Canada: Participatory Research Network, International Council for Adult Education, 29 Prince Arthur Avenue, Ontario, M5R1B2, Canada. The African component started in Tanzania and has spread into a number of African countries. In 1986, Deryk Malenga of the University of Zambia became the President of the African Participatory Research Network.

[7]These categories are adapted from an analysis of the factors likely to influence a role occupant's behavior in response to new norms embodied in law. See Robert B. Seidman, *Law and Development,* (London: Croom-Helm, 1978). See also W.J. Chambliss and R.B. Seidman, *Law, Order and Power* (Reading, Mass.: Addison-Wesley Publishing Co., 1971).

[8]Unlike mainstream economic theory, which focuses on interest as the single most important influence affecting behavior, this approach includes interest along with six other categories.

[9]The governments of Tanzania, Zambia and Zimbabwe have all initiated programs to help ensure that women have the opportunity to participate fully in development projects.

[10]*Cf.* Sherry Arnstein, "Eight Rungs on the Ladder of Citizen Participation," in Edgar S. Calne and Barry A. Passelt, eds., *Citizens Participation: Effecting Community Change* (New York: Praeger, 1971); and Robert K. Yin and Douglas Yates, *Street-level Governments* (Lexington, Mass.: Lexington Books/D.C. Health, 1975. pp. 26-27; see Lisa Peattie, "Participation" in Developing Countries: A Peruvian Case (Atlanta: unpublished paper presented at Collegiate Schools of Planning conference, Nov. 2, 1985) for criticism of the ladder concept as simplistic.

[11]Mahmood Mamdani, "Disaster Prevention: Defining the Problem," *Monthly Review* (New York: October 1, 1985).

[12]Edward C. Banfield, "Ends and Means in Planning," in Andreas Faludi, *Planning Theory* (New York: Pergamon Press, 1973).

[13]*E.g.:* The World Bank evaluation of an El Salvador site-and-service housing project established indicators of housing standards, family satisfaction, and economic value, measured against a pre-determined scale. (See Michael Bamberger, Edgardo Gonzalez-Polio, and Umnuay Sae-Han, Evaluation of Sites and Service Projects: The Evidence from El Salvador, World Bank Staff Working Papers #549, Washington, D.C., 1982).

[14]Essentially positivism.

[15]This view of ends-means capsizes the conventional argument that problem-solving—usually equated with pragmatism—encompasses only incremental change; whereas ends-means alone permits radical change. Introduction of a problem-solving approach within the context of a broader analysis of the causes of underdevelopment, however (see Chapter 3, below), fosters consciousness-raising concerning the need for more basic change.

[16]Karl Deutsch, "Social Mobilization and Political Development," American Political Science Review, Vol. 55, p. 493.

[17]Many who emphasize learning-by-doing, or praxis—ranging from John Dewey, through Paulo Freire to Karl Popper, Sartre and Karl Marx—have contributed to developing this approach. (See R.L. Bernstein, *Praxis and Action.* Philadelphia: University of Pennsylvania Press, 1971).

[18]By incorporating the problem-solving methodology, this approach attempts to further systematize the first of the three basic steps identified by David Korten as central to a participatory approach to development: i) embracing (not rejecting) error; ii) planning with the people; and iii) linking knowledge building with action. ("Community Organizing," *op. cit.*)

[19]Brian H. Smith found similar results (see his U.S. and Canadian Nonprofit Organizations (PVO's) as Transnational Development Institutions. New Haven,

Conn.: Program on Non-Profit Organizations, Institute for Social and Policy Studies, PONPO Working Paper, 70, and ISPS Working Paper, 2070, 1983).

[20]Brian Smith argues that trust is an important element in institutions and network building which thorough evaluations, especially by North Americans, can undermine (Smith, U.S. and Canadian Nonprofit Organizations, *op. cit.)*

[21]Keynote speaker at the Northeast Regional Meeting, National Council for International Health, Park Plaza Hotel, Boston, Oct. 18, 1985.

[22]Except Mozambique and Angola, which inherited illiteracy rates of 90 to 95% from prolonged rule by the Portuguese—one of the most underdeveloped European countries.

[23]The Bagamoyo project in Tanzania supports this proposition; see Marja-Liisa Swantz and Helena Jerman, Bagomoyo Research Project "Jipemoyo;" Introduction to its general aims and approach (Dar es Salaam: Ministry of National Culture and Youth, 1977).

[24]See Ann Seidman, Planning for Development in SubSaharan Africa (New York: Praeger; and Tanzania: Tanzanian Publishing House, 1972), for a further analysis of alternative possible national development strategies. The 1980s crisis that engulfed African and other third world rural populations bears witness to the importance of these issues.

[25]Strengthening national research capacity was also an aim of the Bagamoyo project, Swantz and Jerman, "Jipemoyo," *op. cit.*

A peasant outgrower cotton project. (Photo by Michael Scott, Oxfam America)

LOCATION OF LEARNING PROCESS PROJECTS IN SOUTHERN AFRICA

Map by Jerry Alexander

● Represents approximate location of projects.

CHAPTER THREE
The Three Countries

I didn't realize we faced so many of the same problems. We can learn a lot from each other.
—A project representative at the Lusaka regional workshop July, 1984.

The previous chapter emphasized that the impact of aid depends in part on how, given the constraints on their activities, as well as their resources, the project members behave. That implies that the institutions and attitudes of the community will likely influence their use of aid. A brief review of the socio-economic backgrounds of the three southern African countries where the pilot Learning Process took place may illuminate the implications of its findings.

Before launching into a description of the manifold complexities of the country backgrounds, however, it is necessary to make explicit the underlying values of the exercise,[1] and incorporate them into an analytical framework. The problem-solving methodology requires the researcher to make choices at every step: Whose problems to examine? What explanations to consider? What evidence to gather? What solution to propose? What criteria to use in evaluating them?

Unless researchers consciously explicate their analytical framework, their implicit, unstated values will control those discretionary choices. Insofar as possible, therefore, researchers should make their underlying values or assumptions explicit, formulating them as testable propositions and incorporating them into a logically consistent analytical framework. In the course of using that framework to examine particular problems in the real world, they will also, at least partially, test the validity of the propositions which constitute its building blocks. Over time, they will undoubtedly discover they must revise the analytical framework, or at least some of the propositions of which it is composed, to ensure that it coincides more closely with ever-changing reality. Thus, systematic learning-through-doing facilitates the use of experience, not only to explain and offer better solutions for specific difficulties, but also to test and improve the researchers' analytical framework.

The stated aims of private voluntary organizations engaged in the pilot Learning Process focused the analysis on the problems of the rural poor. The analytical framework, therefore, must describe those features of the countries' socio-economic backgrounds relevant to the rural poor and the difficulties of their daily lives. That framework influenced the Learning Process participants' choice of explanations to consider in the Learning Process, and the kinds of data required to test them. It suggested which proposed strategies to implement, as well as the criteria to use in assessing the consequences.

This chapter:

- delineates the analytical framework used to locate the projects included in the Learning Process in relation to their surrounding communities; and
- employs that analytical framework to outline the evidence revealing the relevant socio-economic background of the projects in the three southern African countries.

THE ANALYTICAL FRAMEWORK

The analytical framework explicating the location of the kinds of projects included in the Learning Process in relation to their surrounding communities may be summarized as follows:

A dualistic resource allocation pattern characterizes the independent southern African states where the Learning Process took place. A century of colonial rule imposed a so-called "modern" sector characterized by foreign-owned mines, settler-owned farms, and a limited number of import-substitution industries. These geared the national economy to the export of crude materials to the factories of Europe and the United States, and the import of manufactured goods—machinery and equipment for the modern sector, and luxury consumer goods for the few who could afford to buy them.

The colonial state shaped the country's institutional structures to provide the low-cost labor which made the colonial enterprises profitable: It enforced laws to push the Africans off the best land, restricting them to the least fertile, most arid areas. It imposed taxes to require them to earn cash. Colonial banks refused to provide African peasants credit to buy needed farm inputs. In some countries, colonial markets discriminated against

African sales of cash crops. As a result, African men had little choice but to migrate to work for wages barely sufficient to support themselves. Their families remained in neglected hinterlands, supposedly supporting themselves, to provide the next generation of laborers. The loss of male labor left women, children and old folk to raise food crops with outmoded tools on "reserved" lands typically characterized by less desirable soils, vulnerable to drought, and boasting few, if any, roads, schools, and clinics.

As a consequence of this pattern of development, at independence the typical southern African state inherited a stratified society with a sharply skewed income distribution. Foreign firms remitted home in the form of interest, dividends and high rates of profit (often concealed by transfer pricing)[2] a significant share of the nation's locally-generated investable surplus. Less than ten percent of the population, predominantly colonial settlers and the managerial personnel of the transnational corporate affiliates, received 50 to 75 percent of the remaining national income. A relatively few wage workers, typically earning less cash than they needed to support their families, lived and worked on settler estates or in urban townships and mining compounds. The bulk of the African population, especially women, children and old folk, barely survived, chronically on the verge of hunger. Many of the children never attended schools. Most had never had the opportunity to visit a clinic. Their life expectancy averaged about a third less than that of citizens of developed countries.

In these remote, historically neglected rural backlands, donor agencies transfer additional resources to projects seeking to enable recipients to improve their living and working conditions.

The Southern African Pilot Learning Process involved the members of a sample set of these kinds of projects in Tanzania, Zambia and Zimbabwe in a participatory process to evaluate their efforts.

Figure 3.1 presents a simple model to help explicate this analytical framework to show the location of the kinds of projects included in the Learning Process in the global system.

UNDERDEVELOPMENT AND POVERTY IN SOUTHERN AFRICA

In the 19th century "scramble for Africa" the squabbling colonial powers ignored existing geographic, economic, and cultural realities. Despite significantly differing social and cultural traditions and national struggles for liberation, the subsequent distorted dualistic growth created remarkably similar rural environments within which the pilot Learning Process participants lived and worked in the early 1980s.

IN THE GLOBAL SYSTEM:

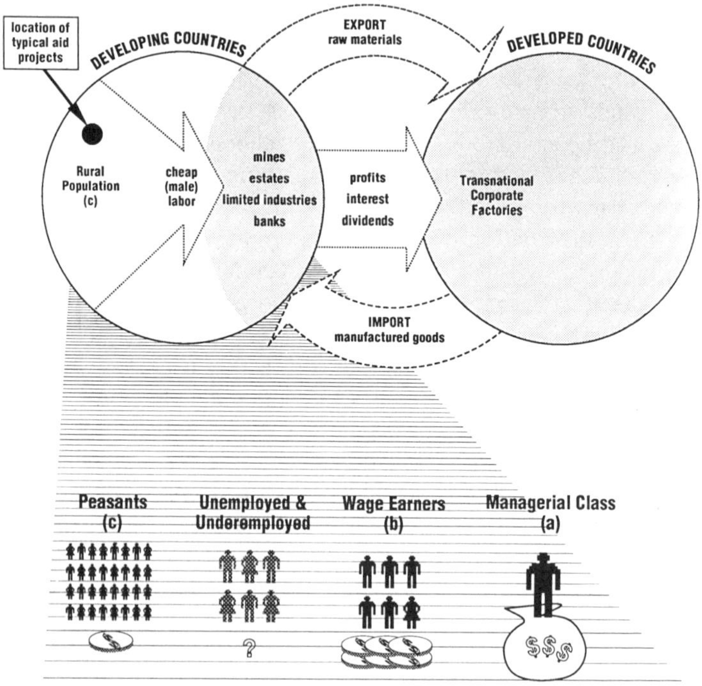

FIGURE 3.1: Model of Underdevelopment and Location of Projects included in the Pilot Learning Process.

Graphics by William Z. Harris

Notes: (a) Managerial, supervisor personnel, a few entrepreneurs, and large commercial farmers; (b) Low-paid wage earners; (c) Because of migratory labor system, in most cases (but not all) women, children and old folk predominate in the rural population.

TANZANIA[3]

Date of independence: 1961

Area: Mainland: 362,688 sq. mi (939,362 sq. km.); Zanzibar and Pemba: 1,020 sq. mi. (2,641 sq. km.)

Population (1983): 19.8 million (plus about 500,000 living in Zanzibar and Pemba)

Life expectancy: 52 years

Percent literacy among adults: Males, 78%; females, 70%

Percent of population urban: 10%

Gross Domestic Product (1981): $4.5 billion

Per capita income (1981): $228
Main exports: Coffee (34%); cotton (16%); cloves, almost entirely grown on Zanzibar (10%); diamonds (6%)
Manufacturing, as percent of GDP: 9%[4]

In 1961, following a relatively brief nationalist campaign, Tanzania became the first of the three countries to win independence. The British always viewed it as something of a colonial step-child. After World War I, they had taken over the Germans' former colony, Tanganyika, as a League of Nations protectorate.[5] New European settlers, including a number of Greek entrepreneurs, had taken over the former German estates. But the British still considered Kenya as the regional sub-center of their East African colonies. They combined Tanganyika together with Kenya and Uganda into the East African Common Market which fostered Kenyan industrial growth at the expense of the other two countries.

At independence, therefore, despite its large territory and access to the sea, Tanganyika remained a predominantly agricultural exporting country with one of the lowest per capita incomes on the continent. Over 100,000 African laborers worked on less than 100 foreign-owned estates to grow the country's main export, sisal. By the early 1960s, however, synthetics had destroyed the world market for natural fibers; world sisal prices plummeted. A few settler-owned coffee estates, together with peasants cultivating the rich soils at the foot of Mount Kilimanjaro, grew coffee. In the region spreading south from the shores of Lake Victoria, Tanganyikan peasant families cultivated cotton for export. Low-cost labor (including family labor)[6] enabled Tanganyikan coffee along with that produced in other African countries, to penetrate the European and American instant coffee markets.

In the western districts, colonial neglect, coupled with the requirement that every Tanganyikan pay taxes, coerced Tanganyikan men to migrate hundreds of miles from their homes to work as low-paid wage labor on the sisal estates. Some trekked across the border to more prosperous settler-ruled farming and mining regions further south. In remote areas like Kigoma,[7] the site of some projects that participated in the Learning Process, most women and old folk stayed home to care for the children and cultivate food crops. At independence, they remained among the poorest districts in the country, with per capita incomes about two-thirds that of the coffee growing areas.[8]

After independence, the new government initially followed World Bank advice to expand export crops and adopted policies seeking to attract foreign investment to build import-substitution industry. Its peasants doubled and even tripled crop exports, including new ones like tea and

TABLE 3.1. Total Foreign Aid Received By Tanzania, 1946-1984 (in US $ millions)

	1946-61	1962-80	1981	1982	1983	1984	Total
US Total Aid	4.4	271.8	37.2	19.6	7.9	6.4	344.1
total grants	2.5	208.8	17.9	14.6	2.9	6.4	249.2
total loans[a]	1.9	63.8	19.3	5.0	5.0	0.0	94.0
US grants:							
AID	0.7	106.2	9.9	10.6	0.0	1.2	125.4
Food for Peace	1.8	91.8	7.3	2.6	1.5	3.9	108.9
Peace Corps		10.0	0.7	1.4	1.4	1.3	14.9
Military	—	—	—	—	—	—	—
US Loans:							
AID	1.9	30.6	11.8	—	—	—	42.1
Food for Peace	—	33.2	7.5	5.0	5.0	—	51.9
Other loans	—	—	—	—	—	—	—
Other US:	—	14.1	1.7	—	—	—	16.0
Ex-Im Bank	—	13.8	1.7	—	—	—	15.7
Other	—	0.3	—	—	—	—	0.3
Multilateral aid:							
Total:	3.2	1014.1	107.2	11.7	74.2	38.9	1288.4
IBRD	—	318.2	—	—	—	—	318.2
IFC	2.1	5.2	—	—	—	3.9	11.0
IDA	—	550.7	92.8	75.0	46.8	35.0	799.9
AFDB	—	53.5	13.8	8.6	22.4	—	97.9
UNDP	0.9	54.1	0.6	14.7	5.0	—	75.3[b]
Other UN	0.2	27.4	—	13.4	—	—	41.0
EEC	—	5.0	—	—	—	—	5.0

TABLE 3.1. *Continued.*

	1978-81	1981	1982	Total
Other Western aid:				
Total:	1,565.1	493.9	473.1	2,532.1
United Kingdom	140.8	56.0	41.2	238.0
Sweden	285.9	76.5	73.8	436.2
West Germany	345.8	54.5	59.0	459.3
Netherlands	263.7	74.4	57.5	395.6
Other	586.9	199.5	216.6	1,003.0
OPEC	—	6.5	—	6.5

	1954-81	1982	Total
Socialist countries			
Total reported:	485	15	500
USSR	40	5	45
Eastern Europe	75	10	85
China	370	—	370

Notes:
[a]Tanzania repaid $23.7 million of these loans, $17.1 million to AID, $6.6 million to Food for Peace, and $6.5 million to the Export Import Bank.
[b]The original source reports a total of $15.4 million; the figure given here is the sum of the reported yearly aid.

Source: US Agency for International Development (AID), Congressional Presentation, Annex I, Africa, Fiscal Years, 1985 and 1986 (Washington, D.C.: AID, 1984 and 1985) Table for each country.

tobacco. After several years, however, Tanzania discovered that, because of worsening terms of trade, its foreign exchange earnings failed to expand and foreign manufacturing investment did not materialize. Therefore, in 1967 the ruling party announced at Arusha that it would begin to take over the national "commanding heights"—banks, export-import trade, and basic industry—explicitly in order to redirect the nation's development along a path of more self-reliant development. Unfortunately, however, the government failed to develop a long-term strategy to achieve balanced industrial and agricultural growth to increase productive employment, raise living standards, and reduce dependence on crude agricultural exports.[9]

Meanwhile, President Nyerere and the ruling party urged peasants throughout the country to move voluntarily from their scattered individual family homesteads into *ujamaa* villages—a kind of family-based socialism—in order to benefit from improved social infrastructure: schools, clinics, community bore-holes. The party leadership hoped *ujamaa* villagers would pool their resources to acquire improved technologies and increase productivity on communal land holdings. This, they maintained, would enable the villagers to finance the new social facilities.

In the early 1970s, the party sought to hasten families' movement into villages. In some places, the implementation of this policy gave rise to criticisms of undue use of coercion. By the end of the decade, most families had joined new *ujamaa* villages. In 1972, for example, "Operation Kigoma" involved planning teams which mapped out sites to disperse highland overpopulated settlements, and consolidate dispersed *miombo* lowland zone settlements. By 1974, 95 percent of the Kigoma rural population was consolidated into 193 registered villages.[10] In the past, slash-and-burn agriculture, hoe cultivation and flimsy houses that people could easily move to new sites had characterized the scattered settlements of the *miombo* lowlands of Kigoma. After *ujamaaization,* sturdy mud and brick houses lined planned village roads in communities of 200 to 500 families. Every ten families elected a representative to village councils in what became known as the ten-house cell system.

Critics argued, however, that planners had made mistakes in locating the new villages sometimes quite far from previously cultivated and fertile areas.[11] In some cases, the relocation process interfered with on-going cultivation activities.

The accelerated *ujamaaization* took place in the midst of the drought of the early 1970s, which aggravated whatever shortages of food and agricultural exports may have resulted from disturbing the normal cropping cycle. In the late 1970s, the invasion by Idi Amin's troops and the prolonged war with Uganda imposed further financial burdens on the nation, including heavy spending of foreign exchange to import military equipment. The war itself disrupted agricultural production in the northern border areas, especially affecting coffee cultivation. In other areas, the enrollment of youth in the army reduced the available labor force. Increased expenditures on the armed forces and rural social services, not financed by expanded productivity and rising government revenues, forced the government to borrow heavily. Its internal debt fostered inflationary pressures which, accelerated by rising oil prices,[12] spurred rapidly rising domestic living costs. All these factors aggravated what appeared as the negative impact of *ujamaaization* on rural productivity and living standards. Nevertheless, in spite of the drawbacks of the accelerated *ujamaa* village program, by the mid-1980s when the pilot Learning Process took place, increasing numbers of people argued that few villagers would give up their new communities to return to their separate homes.

The Tanzanian government's failure to formulate and implement a strategy to restructure the national economy to reduce its dependence on crude agricultural exports, however, left the national economy still extremely dependent on the world market.[13] When the international recession worsened Tanzania's terms of trade, it could no longer afford to pay its heavy international debts and continue to import the machinery and equipment needed for transport and increased productivity. By 1982, the World Bank had become Tanzania's largest creditor—$792 million.[14]

The government turned to the International Monetary Fund for assistance, but rejected that agency's conditions. Government spokespersons argued the conditions would aggravate unemployment and reduce social services, shifting the burden of the crisis to the poor who could least afford to bear it.

Over the next years, the government sought to implement its own structural adjustment program.[15] It partially met the IMF's demands by reducing expenditures and laying off government employees. Without the IMF's seal of approval, most private Western banks and some Western governments (notably the United States) refused to extend further credit to meet the nation's foreign exchange requirements.[16]

ZAMBIA

Date of independence: 1964
Area: 290,724 sq. mi. (752,975 sq. km.), landlocked
Population (1983): 6 million
Life expectancy (1983): 50 years
Percent literacy among adults (1983): Males, 79%; females, 58%
Percent of population urban (1983): 40%

Gross Domestic Product (1981): $3.8 billion
Per capita income (1981): $638
Main exports: Copper (85%); cobalt (10%)
Manufacturing, as percent of GDP: 19%; employment: 45,510[17]

Early in the 20th century two foreign mining companies, one based on South Africa and the other in the United States, developed four mines in Zambia (formerly known as Northern Rhodesia) to extract copper from the rich deposits of what became known as the Copper Belt. To produce food for the copper miners, the colonial state turned over to white settler farmers the lands for 20 miles on both sides of the railroad stretching from the Copper Belt to the Zimbabwe border (from there, the railroad freighted Zambian copper to South African ports). The colonial state imposed taxes and agricultural marketing and credit facilities to coerce African men to migrate to work on the mines and estates for wages barely sufficient to support themselves.[18]

By independence, some 40% to 60% of the men between the ages of 20 and 40 migrated annually from villages searching for wage employment. By 1980, although the international recession as well as investment in capital-intensive machinery had reduced their numbers, the workers on the mines still produced over three-fourths of the nation's exports.[19]

The loss of male labor from the rural areas destroyed previously complex labor-intensive farming systems, some built around extensive irrigation canals, others requiring constant clearing of new lands to permit old ones to lie fallow to recuperate. Left behind in remote rural villages, almost entirely neglected by the colonial state, women, children and old folk used outmoded tools to scratch a living out of sandy, poorly-watered, often over-used soils.

In 1964, Zambia's nationalist movement, led by the mine workers, won independence relatively easily. The new government declared its "humanist" goal of closing the rural-urban gap. It multiplied expenditures for education, health facilities, and roads in remote rural areas. It expanded the state bureaucracy, fostering the emergence of what some critics termed a "bureaucratic bourgeoisie." At the same time, it extended state intervention into the mining business in partnership with the big foreign mining firms and expanded the parastatal sector, creating an additional managerial elite.[20] To reduce its dependence on then-minority-ruled Rhodesia (now Zimbabwe) and South Africa, the government invested heavily in new routes to the coast through Tanzania. By cutting off the trade with the minority-ruled states to the South, it created a protected market which—with state support—briefly attracted foreign investment in import-substitution industries.[21]

TABLE 3.2. Total Foreign Aid Received by Zambia, 1946-1984 (in US $ millions)

	1946-61	1962-80	1981	1982	1983	1984	Total
US Aid: Total:	27.7	157.5	30.6	27.1	27.1	36.4	305.4
Total Grants:	0.3	24.5	5.6	5.1	5.9	26.4	66.8
Total Loans[a]:	27.4	133.0	25.0	22.0	22.0	10.0	238.6
US Grants:							
AID:	0.3	10.3	5.6	5.1	0.5	21.3	42.1
Food for Peace	—	14.2	—	—	5.4	5.1	24.7
Other	—	—	—	—	—	—	—
US Loans[a]:							
AID	5.0	95.0	15.0	15.0	15.0	—	144.2
Food for Peace	—	38.0	10.0	7.0	7.0	10.0	72.0
Other	22.4	—	—	—	—	—	22.4
Multilateral aid							
Total:	64.3	652.8	46.1	140.2	40.1	104.7	1,028.5
IBRD	63.5	498.6	26.0	11.7	—	75.0	655.6
IFC	—	41.5	—	34.4	18.8	5.8	100.4
IDA	—	37.4	—	50.5	20.3	22.4	130.5
AFDB	—	43.4	9.7	9.5	—	—	62.2
UNDP	0.8	27.7	0.9	2.6	1.0	—	32.5
Other UN	—	2.1	1.5	—	—	1.5	5.3
EEC	—	2.5	8.0	31.5	—	—	42.0
	1978-1981	1982	1983	Total			

(Continued)

TABLE 3.2. *Continued*

Other Western aid:				
Total:	653.8	169.6	157.4	980.8
United Kingdom	169.3	22.5	21.1	212.9
Sweden	103.2	27.5	29.4	160.1
West Germany	93.8	29.0	25.0	147.8
OPEC	—	11.5	—	11.5

	1954-82	1982	1983	Total
Socialist Countries				
Total reported aid:	524	—	17	541
USSR	21	—	9	30
Eastern Europe	166	—	—	166
China	337	—	8	345

Note: [a]Zambia repaid $47.7 million of the loans: $16.8 million to AID, $3.8 million to Food for Peace, and $27.1 million other loans.

Source: See source for Table 3.1 above.

But these measures left largely intact the inherited state structure, particularly the institutions governing the allocation of resources. They failed to capture the locally-generated surpluses for investment in industry and agriculture to create a more balanced, integrated and self-reliant national economy. Many young women and men, learning in new schools of the alternatives to stagnant rural poverty, fled to squatter compounds that skirted cities on the Copper Belt and along the railroads. In ten years, the urban population jumped from 25% to 40% of the nation's inhabitants. But the lives of those peasant families who remained in the scattered rural hamlets changed little. Zambia became increasingly dependent on imported foodstuffs to feed its cities' growing population.[22]

In the 1970s, as Zambia and other copper exporting countries expanded their output, the world copper price plummeted. Zambia's export earnings and government revenues shrank. The nation had to borrow heavily to finance its expanded social and economic infrastructure and to pay for the imports, including food, on which the economy depended. By the 1980s, Zambia was spending almost two-thirds of its export earnings simply to service its mounting external debt. Shortages of materials, spare parts and equipment forced its industrial sector to reduce output and lay off workers, and hindered transport into remote rural areas.

The government requested the International Monetary Fund for assistance, receiving about a third of IMF's total credit to the region. In return, it adopted IMF austerity measures and began to stress increased market-oriented policies. It reduced government spending, laid off government workers, increased the bank rate, raised food prices and gradually ended subsidies, and devalued the *kwacha.* In 1985, it began to auction its foreign exchange, leading to a drastic decline in the value of the *kwacha* and price increases throughout the economy.[23]

ZIMBABWE

Date of independence: 1980
Area: 150,866 sq. mi. (390,759), landlocked
Population (1983): 7.5 million
Life expectancy (1983): 53 years
Percent literacy among adults (1983): Males, 76%; females, 71%
Percent of population urban: 20%
Gross Domestic Product (1981): $5.9 billion
Per capita income (1981): $786
Main exports: Tobacco (28%); gold (9%); iron products (15%); sugar (7%); nickel (5%)
Manufacturing, as percent of GDP: 25%; employment, 180,500

TABLE 3.3. Foreign Aid to Zimbabwe, 1946-1984 (in US $ millions)

	1946-61	1962-80	1981	1982	1983	1984	Total
US Total	5.6	24.5	27.8	75.0	64.0	47.8	244.5
Total grants:	0.6	24.5	27.8	75.0	64.0	47.8	239.5
Total loans[a]:	5.0	—	—	—	—	—	5.0
US Grants:							
AID	0.6	24.5	25.0	75.0	60.0	41.0	230.9
Food for Peace	—	—	2.8	—	4.0	6.8	13.6
Military	—	—	—	0.1	0.1	0.2	0.4
US Loans:							
AID	5.0	—	—	—	—	—	5.0
Ex-Im Bank	—	—	33.3	6.1	—	—	39.4
Multilateral aid							
Total:	83.3	11.1	149.4	14.0	267.6	113.4	638.6
IBRD	83.1	3.9	92.0	—	202.0	96.1	477.1
IFC	—	—	38.0	—	—	2.3	40.3
IDA	—	—	15.0	—	38.9	—	53.9
AFDB	—	—	—	12.2	—	—	12.2
UNDP	0.1	4.2	1.3	1.8	2.4	—	9.7
Other UN	0.1	3.0	3.1	—	4.3	—	10.4
EEC	—	—	—	—	20.0	15.0	35.0
	1978-81	1982	1983	Total			

Other western countries-				
Total reported:	254.2	135.4	128.4	518.0
United Kingdom	133.5	37.6	22.1	193.2
West Germany	32.0	23.0	34.1	89.1
Sweden	22.5	10.1	15.3	47.9
Netherlands	16.3	14.3	9.6	40.2
Other	49.9	50.4	47.3	147.6

	1954-81	1982	1983	Total
Socialist countries				
Total reported:	11	2	30	43
USSR	—	—	—	—
Eastern Europe	—	2	30	32
China	11	—	—	11

Note: [a]Zimbabwe repaid US $7.7 million, of which $6.6 went to the Export Import Bank, leaving an overpayment of $2.7 million unexplained by the source.

Source: See Source for Table 3.1.

Zimbabwe's environment differed from Zambia's and Tanzania's primarily in the form of colonial rule and the liberation struggle. In the early 20th century, concluding that Zimbabwe (then Southern Rhodesia) lacked major mineral deposits, the colonialists turned over half the nation's land—the most fertile, well-watered areas—to white settler farmers to cultivate crops, primarily for export. The settlers rapidly became the major influence in the colonial government. They exercised state power to push the formerly prosperous African peasants into reserves misnamed "Tribal Trust Lands." As in Zambia, the colonialists imposed taxes and discriminatory market and credit institutions to force African males to migrate to work for low wages on the white-owned farms and mines.

In 1953, the colonial settlers succeeded in uniting Zimbabwe (then Southern Rhodesia) with Zambia (then Northern Rhodesia) and Malawi (then Nyasaland) in the Federation of Rhodesia and Nyasaland. Dominating the Federation, the Southern Rhodesian settlers exercised their control of the federal government to tax the Zambian mines to finance the advanced economic infrastructure required to attract foreign investment to build up their industry; and to encourage the migration of Malawian workers to keep down wages on their mines and farms. In 1963, Zambian and Malawian nationalists, calling for independence and majority rule, demanded an end to the Federation.[25]

In the 1960s, about 6000 white commercial farms, spread over the best half of the nation's land and employing over 300,000 (mostly male) workers, produced 70 percent of the country's marketed crops and 90 percent of its agricultural exports. The state power financed expanded economic infrastructure and subsidies to stimulate these commercial farms' output to maintain national food self-sufficiency as well as to expand export crops.

Meanwhile, the remnants of some 750,000 peasant families—mainly women, children and old men—struggled to survive in the overcrowded reserves. Even the colonial authorities admitted that the Tribal Trust Lands could not realistically support half that number. The per capita food production of the peasant farmers in those areas declined.[26]

After Zambia and neighboring Malawi attained majority rule, the white minority in Zimbabwe—about three percent of the population—unilaterally declared independence (UDI). Despite United Nations sanctions, the minority continued to rule the country for a decade and half. The Zimbabwean liberation movement mobilized the population to support guerrilla warfare to win the right to vote and restoration of their land. In the late 1970s, the increasingly effective armed struggle rendered much of the countryside ungovernable. Foreign capital began to flee.

In 1980, at Lancaster House in London, the Zimbabwean liberation movement, under American and British pressure, agreed to a compromise constitution.[27] In the ensuing election, the black majority gave 57 percent of the vote to the Zimbabwe African National Union—Patriotic Front (ZANU-PF), the party of Robert Mugabe. He became the new nation's first Prime Minister.

Like the governments of Zambia and Tanzania, the new Zimbabwe government multiplied expenditures on schools, health and roads. It created new local government institutions, district councils run by Africans, in the former Tribal Trust Lands, now renamed "Communal Areas." It also began to resettle some of the landless peasants on abandoned farms purchased from white settlers. Although half the commercial lands remained un- or underutilized, the majority of African peasants—some 800,000 families—still struggled to survive on the overcrowded, infertile Communal Areas. During the three year drought (1981-83), some 40 percent lived primarily on drought rations provided by the government with foreign assistance.[28]

SUMMARY

Since the differing socio-economic contexts of the three countries may differentially affect project members' utilization of aid, a knowledge of the communities where the pilot Learning Process took place may illuminate its findings. Private voluntary organizations participating in the pilot Learning Process aimed to provide assistance to projects belonging to the rural majority. The analytical framework underlying the Learning Process helps to locate these kinds of projects in the world system. Thus it provides a guide for describing the relevant characteristics of the three southern African countries—Tanzania, Zambia and Zimbabwe—which influenced the Learning Process findings. Examination of the findings should contribute to testing not only the causal explanations offered for project success or failure, but also the model purporting to explicate the differing socio-economic backgrounds in which the rural poor live.

CHAPTER THREE
Notes

[1]Alvin Gouldner, in *The Coming Crisis of Western Sociology* (New York: Avon, 1971), called these "domain assumptions." Chapter 2, above, includes them in the category, "Ideology."

[2]That is, the firms raised the prices of imported machinery, equipment and materials above the prices prevailing in world markets for those items, and reduced the prices of exported crude materials below the relevant world prices. Thus the profits appeared on the books, not of their affiliate in the Third World country, but their overseas affiliate. (For further discussion, see Robin Murray, *Multinationals Beyond the Market.* New York: Wiley, 1981).

[3]The editors wish to express appreciation to Anita Baltherzen, who, as part of the Applied Development Research Network, conducted background research for this section (see her MA thesis, Clark University, International Development and Social Change Program, 1986).

[4]A. Seidman, *The Roots of Crisis in Southern Africa* (Trenton, N.J.: Africa World Press, 1985).

[5]The colonists used the name, Tanganyika, for what is now mainland Tanzania: but when after independence the rich clove-growing islands of Zanzibar joined it, the federated state became known as Tanzania.

[6]Most peasant families grew their own foodstuffs, thus reducing the monetary remuneration required for the family's subsistence from cash crops or wage employment.

[7]The British colonial administrator, Major S.S. Orde Browne, opposed cash cropping in Kigoma (Wayne and Howard, 1975, p. 25), for he envisioned that area as a labor reserve. Colonial administrators received a commission for each man recruited for the sisal estates (Kavura et al, n.d.; and Coulson, 1977).

[8]James Mittelman, *Underdevelopment and the Transition to Socialism: Mozambique and Tanzania* (New York: Academic Press, 1985).

[9]Andrew Coulson, *Tanzania, 1800-1980: A Political Economy* (New York: Oxford University Press, 1982); Barbara Dinham and Colin Hines, *Agribusiness in Africa—A study of the impact of big business on Africa's food and agricultural production* (Trenton, N.J.: Africa World Press 1984).

[10]Jons Oldewelt, Time Utilization of an African Peasantry: A Case Study from Kigoma, *Tanzania* (Denmark: Centre for Development Research, 1984).

[11]For a discussion of *ujama* villagization, see Michael McCall and Margaret Startsch, "Strategies and Contradictions in Tanzania's Rural Development: Which Path for the Peasants?" in Lea and Chaudhri, eds., *Rural Development and the State* (London: Methuen, 1983, pp. 241-272). See also Coulson, Tanzania, op. cit., Cheryl Payer, "Tanzania and the World Bank," *Third World Quarterly,* 5, 4 (October, 1983); P.L. Raikes, "Ujamaa and Rural Socialism," *Review of African Political Economy,* 3 (May-October, 1975).

[12]By 1979, oil consumed over half of the nation's foreign exchange earnings (See Ellen E. Hanak, *The Tanzanian Balance of Payments Crisis: Causes, Consequences and Lessons for a Survival Strategy* (Dar es Salaam: University, ERB Paper 82.1, 1982).

[13]Several authors suggest the World Bank played a major role in influencing Tanzania to focus on expanding agricultural exports; e.g. Coulson, Tanzania, *op. cit.;* Theresa Hayter and Catherine Watson, *Aid: Rhetoric and Reality* (London: Pluto Press Ltd, 1985; Goran Hyden, *Beyond Ujamaa in Tanzania: Underdevelop-*

ment and an Uncaptured Peasantry (Berkeley: University of California Press, 1980); Frances Moore Lappé and Adele Beccar-Varela, *Mozambique and Tanzania: Asking the Big Questions* (San Francisco Institute for Food and Development Policy, 1980).

[14]Economist Intelligence Unit, Annual Summary, 1985, p. 21.

[15]*Op. cit.,* 1983.

[16]African Research Bulletin, 20, #6, 6910.

[17]Seidman, *op. cit.* p. 153.

[18]Lewis Gann, *Central Africa: The Former British States* (Englewood Cliffs, N.J.: Prentice Hall, 1971).

[19]Republic of Zambia, Monthly Digest of Statistics, Vol. XVII, Nos. 7-12 July/Dec 1981 (Lusaka: Central Statistical Office, 1982).

[20]Richard Sklar, *Corporate Power in an African State: The Political Impact of Multinational Mining companies in Zambia* (Berkeley: University of California Press, 1975).

[21]A. Seidman, "Import Substitution Industry in Zambia," in Ben Turok, *Development in Zambia: A Reader* (London: Zed Press, 1979).

[22]Michael Bratton, *The Local Politics of Rural Development: Peasant, Party and State in Zambia.* (Hanover, NH: University Press of New England, 1980).

[23]Neva Malzgetla, "Theoretical & Practical Implications of IMF Conditionality in Zambia," *Journal of Modern African Studies,* September, 1986.

[24]Seidman, *op. cit.,* p. 155.

[25]J. Deary, "Break-Up: Some Economic Consequences for the Rhodesias and Nyasaland," The Central African Examiner, July, 1963.

[26]Cf. Roger Riddell, "From Rhodesia to Zimbabwe: The Land Question" (London: Catholic Institute for International Relations, 1980.

[27]Jeffrey Davidow, *A Peace in Southern Africa: The Lancaster House Conference on Rhodesia, 1979* (Boulder, Colo.: Westview Press, 1984); and David Martin and Phyllis Johnson, *The Struggle for Zimbabwe: The Chimurenga War* (Harare: Zimbabwe Publishing House, 1981).

[28]D. Weiner, S. Moyo, B. Munslow, and P. O'Keefe, "Land Use and Agricultural Productivity in Zimbabwe," Journal of Modern African Studies, Vol. 23, No. 2, pp. 251-285.

Working on a community construction project. (Photo by Oxfam America staff)

Poultry club members. (Photo by Oxfam America staff member)

CHAPTER FOUR
Learning By Doing

This is a different kind of research than I have ever taken part in before.
—A researcher at the Lusaka Workshop, July 1984.

The underlying theory of the pilot Learning Process posits that the members, themselves, could learn best about what kind of aid "works"—and what does not—through participating in evaluating their own projects.[1] In the course of planning and implementing the process, project members, intermediary and donor agency staff, and national researchers could systematically improve their knowledge of the causes of the difficulties plaguing their activities and discover better ways of attaining their goals.

This chapter describes the implementation of the pilot Learning Process. In particular, it discusses:

- the criteria used in selecting the projects and the participants for the pilot Learning Process;
- the projects included;
- the way in which the participants designed the pilot process at the first regional workshop in Lusaka, Zambia;
- the steps taken in implementing the pilot process; and
- the structure and function of the second regional workshop at Gwebi in Zimbabwe.

THE SELECTION OF THE PROJECTS

The initial planning for the Learning Process took place in Boston. Members of the Applied Development Research Network, including several southern Africa graduate students from universities in the Boston area, discussed the kinds of issues and procedures which the pilot project might adopt.[2] An advisory board made many useful suggestions.[3] All these

Box 4-1: The Steps in the Learning Process

I. January-April, 1984: Preparation: Initial discussions in Boston by Oxfam America, advisory board, and other private voluntary organizations.

II. April, 1984: Oxfam America staff member visits southern Africa to discuss proposed Learning Process with intermediary agency staff members and researchers in universities and research institutions in Tanzania, Zambia, Zimbabwe.

III. July, 1984: First regional workshop in Lusaka, Zambia, brings together representatives of projects and intermediary staff; three national researchers from Tanzania, Zambia, and Zimbabwe; and observers from Botswana, Somalia, India, and Dominica to design the Learning Process.

IV. First week of long university vacations in Zambia (August, 1984) and Tanzania (April, 1985): National workshops—including project representatives, intermediary staff, national researchers and students—design national Learning Process.

V. Six to eight weeks during long university vacations in all three countries:[4] students live and work on projects with members to implement Learning Process. Halfway through this period, students meet as group with national researcher to discuss progress and problems.

VI. Final national workshops in Zambia and Tanzania: project representatives, intermediaries, national researcher, and students meet to discuss findings and conclusions.

VII. August, 1985: Final regional workshop in Gwebi, Zimbabwe, where project representatives, intermediary staff, and national researchers from all three countries, plus observers from Dominica and India, meet to share findings, decide to institutionalize the Learning Process in southern Africa.

preliminary discussions, however, emphasized that southern African participants would have to make the final decisions as to the structure of the pilot process.

A few months before initiating the pilot process, an Oxfam America staff member visited Zambia, Zimbabwe and Tanzania. She arranged with a member of African National Congress of South Africa (ANC), who was on leave, to organize the first regional workshop in Lusaka.[5] With the staff of participating intermediary agencies and national research institutions in the three countries, she also discussed the following issues relating to the selection of the individuals who would take part in the project:

- The choice of grassroots projects to participate in the pilot process involved considerable debate, both in the preliminary discussions in Boston and during the selection process in southern Africa. To compare the factors causing success or failure, some people suggested that the pilot project should include some successful and some unsuccessful projects. However, deeper analysis underscored the fact that every project, no matter how successful it appeared, encountered difficulties; and no project completely failed. Hence, success or failure do not constitute very useful criteria. The Tanzanian attempt to employ these criteria in choosing projects illustrated this difficulty: Some of those initially chosen as successful suffered severe setbacks in the course of the year—in the Kigoma maize-bean project (See Table 4-1) the oxen supplied for the "successful" project died within a week of their arrival; whereas some projects initially identified as failures began to resolve their difficulties.

- The first regional workshop agreed that the sample of projects in each country should include some large and some small rural projects engaged in activities of the kind that typically receive aid from private voluntary organizations. As much as possible, they should be located in geographically diverse areas to facilitate analysis of the effects of differing socio-economic contexts within as well as between countries. Unfortunately, because of transport and communications problems, representatives came to the initial Lusaka workshop from only some of the projects that ultimately participated in the learning process.[6]

- From their own staffs, the interested donor and intermediary agencies chose the individuals they thought could contribute to and benefit most from the pilot Learning Process. For the most part, those chosen worked directly with the projects included in the pilot process. The Oxfam America staff members decided to maintain a low profile to enable the southern African participants to take the lead in planning the process. Only one Oxfam America staff member remained throughout the entire first workshop. Two others came only for three days as observers.

- The criteria used to select the national researchers included: a sympathetic interest in analyzing the impact of aid on grassroots projects; knowledge of the relevant information currently being gathered in national and regional teaching and research institutions; an adequate grasp of the necessary research tools; a willingness to learn from as well as give leadership to the national and regional learning process; and ties to a national research/teaching institution which, if the pilot project proved successful, could facilitate institutionalization of a more extensive participatory, problem-solving evaluation process throughout in the region.

- To facilitate the establishment of links with evaluation activities in other Third World regions, Oxfam America invited observers from Central America and India to participate in both the Lusaka and subsequent Zimbabwe workshops.

THE PROJECTS INCLUDED IN THE PILOT LEARNING PROCESS

As finally constituted, the pilot Learning Process included fifteen projects: four in Zambia, five in Zimbabwe, and six in Tanzania. Several private voluntary organizations provided aid to these projects. Oxfam America funded only a few of them alone. Most received aid from several donors, including Oxfam America. In some cases, other private voluntary organizations transferred resources to the projects without any Oxfam America participation.

To avoid any possible negative consequences for any of those who took part in the process, this report follows the ground rule, agreed to by all the participants at the outset, that future publications would not identify participating individuals, projects, or donor agencies by name.

The four Zambian projects

1. A small scale industry near the Mozambique border employing about 40 workers. They manufacture about 50 hoes a day, using scrap metal from old cars and a bridge destroyed during the Zimbabwe liberation war. The workers had built an ingenious home-made blast furnace out of an oil drum, using a bicycle wheel to run the bellows. They also make furniture of all kinds, from school desks to beds. The donor agencies financed the purchase of capital equipment, including a welding machine, a van, and uniforms for the workers. One agency provided a volunteer to assist with the steel-making technology.

In addition to the small industry, the national Zambian researcher, working closely with the women's project officer employed by a local non-government organization to assist women's groups, selected three Zambian women's projects in various stages of development. They sought to test the hypothesis, advanced at the first regional workshop in Lusaka, that project holders should be engaged in participatory assessment of their work from the earliest possible stage. The three women's projects they chose included:

2. A group of women—mostly workers or wives of workers on a coffee plantation in northern Zambia—who sought to obtain funds to initiate income generating activities. They had talked with various donor agencies, but so far had received no outside assistance. Their goals remained ill-defined.

3. About 12 women, members of a church in a rural area on the Copper Belt, who had initially grown vegetables to provide food for the local hospital and needy families. They had also grown some cotton for sale to earn cash income. A donor agency had recently provided them with funds to clear more land and plant more vegetables, both for sale and their own consumption.

4. About 25 women in the western province who had received aid about two years ago to acquire sewing machines to sew school uniforms and protective clothing for council workers. Since the nearest store for purchasing uniforms—required by all school children—was over 50 miles away, the women planned to sell the uniforms to fill a community need. They also hoped to earn money for themselves.

The five Zimbabwean projects

1. A producer cooperative established by ex-freedom fighters and unemployed workers to operate a former commercial farm they had received from the Ministry of Lands and Resettlement. The ex-freedom fighters had contributed their demobilization pay to finance the cooperative's initial expenditures. As new members joined, the cooperative also obtained donations and borrowed funds from a government parastatal, the Agricultural Finance Corporation (AFC), to purchase capital equipment, seed and fertilizers and to cover increased operating costs.

2. One of 10 cooperatives in a northeastern communal area, formed by 200 peasant families to purchase tractors to replace the cattle the peasant families had lost during the liberation struggle when they had been incarcerated in a "protected" village; the peasants maintained that tsetse fly had spread through the region, rendering oxen-drawn plowing no longer viable. A donor agency had provided the down-payments and helped the groups borrow funds from the AFC.

3. A revolving credit scheme which, with the assistance of government-trained women community workers, enabled women's groups to engage in income generating projects in a southern communal area. The donor agency provided the initial funds for the scheme.

4. A training institute established by a non-government community-based umbrella organization in midwest Zimbabwe. This institute aimed to train young men and women, selected by their villages, in skills like brickmaking, carpentry and sewing. The trainees expected to return to their villages to help implement rural development projects.

5. An integrated communal lands development program in central Zimbabwe, involving about 1,500 people in several villages in some 64 projects including such activities as vegetable gardening, sisal/cement roof-making, cattle rearing, uniform sewing, and welding. Women made up about three-fourths of the project participants.[7]

The six Tanzanian projects

In Tanzania, the national researcher and the intermediary agencies, together, selected two projects from each of three community-wide development programs:

1 & 2. Two *ujamaa* villages out of four in Kigoma, western Tanzania, which received funds to purchase six oxen apiece and seed to grow maize and beans on communally-owned acreage. The projects aimed to produce food to generate increased cash incomes for the villagers and food supplies for the nation. The villagers, organized through the 10-house cell system, worked together to prepare the soil and to plant, weed and harvest the crops.[8] Over time, with the assistance of improved technology in the form of ox-drawn plows instead of the traditional hoes, the project aimed to increase the size of the communally-developed acreage.

3 & 4. A timber project and a carpentry project in Ruvuma in southern Tanzania, which provided villagers with tools to cut local timber for

construction and furniture making. By establishing these small industries, the projects aimed to train and employ primary school dropouts, and fulfill a demand for the goods they produced.

5 & 6. As part of a larger food for development program in central Tanzania, a church-related agency sold PL480 foodstuffs (powdered milk, bulgur and cooking oil) at below-market prices to women in the two villages.[9] The funds obtained through these sales were to be used to initiate food-producing activities such as the purchase of goats. The project aimed at improving the nutrition of the children, and encouraging village women to work in groups to develop their own food supplies. The church agency provided courses to teach the women who came to buy the food about the value of adequate nutrition.

THE LUSAKA WORKSHOP

In July, 1984, as a first step in the pilot process, representatives of several intermediary organizations and selected projects in Tanzania, Zambia and Zimbabwe met with national researchers in a learning center about 10 kilometers from the Zambian capitol, Lusaka. There, they worked together for a week to design the year-long process.

As the first stage in the Learning Process, the Lusaka workshop provided a framework within which the participants not only designed the pilot project, but also came to know one another and the kinds of problems they faced. On the first day, people introduced themselves and described their projects. They then elected a steering committee consisting of a representative of a project or intermediary agency and a researcher from each country. The Oxfam America staff member and the workshop organizer participated as *ex officio* members of the steering committee, both to provide information and to make arrangements as needed. The steering committee met each evening to draw up plans for the next day. Its members elected the representative from a Zimbabwe project, the training institute,[10] as chairperson.

The following day, speaking in his local language,[11] the chairperson set the tone of the workshop: A group of southern Africans from differing backgrounds working together to build an effective, self-reliant evaluation process to strengthen their capacity to achieve grassroots development.

After the initial introductions and agreement on the tentative agenda for the week, the workshop participants went to work. Sometimes they divided into small groups with project representatives from all three countries to debate the causes of specific problems and how their assessments could help them to devise better ways of dealing with them. At other times, they

TABLE 4.1. The Projects Included in the Pilot Learning Process—A Summary

Country/project	Project location	Number of members	Activity of Project
ZAMBIA:			
*1. Small-scale industry	Eastern	40	Making and repairing hoes, farm equipment, furniture.
2. Women's group #1	Northern	25	Not well-defined: income-generating.
3. Women's group #2	Copper Belt	12	Vegetable and Maize growing.
4. Women's group #3	Western	25	Sewing uniforms and protective clothing for District Council workers.
ZIMBABWE:			
*1. Producer cooperative	Northeast	40	Commercial farm: coffee, maize
*2. Tractor scheme	Northeast	200 families	Cooperative to buy tractors for cotton farming, etc.
*3. Revolving credit fund	South	80 women's groups	To finance income generating projects for women (gardening, sewing, etc.)
*4. Training institute	Midwest	25	Skills training for communal area development.
5. Integrated communal area program	Central	1,500 members	To stimulate group income-generating activities among communal area inhabitants.
TANZANIA:			
1 & 2. Two village projects	West	about 250 families each	Received oxen, seed to grow beans, maize, on community land.
3. Carpentry project	South	5	To provide local boys tools, training to make furniture.
4. Timber project	South	30	To provide tools, train loggers to cut, prepare local construction timber.
5 & 6. Two village projects	Center	565 mothers, kids in one; 738 in other	Church agency sells PL480 food to women; proceeds to buy inputs to raise goats, grow vegetables, etc.

Note: * = represented at the Lusaka workshop.

met in national groups to discuss how to implement specific aspects of the methodology in their own countries.

The steering committee also planned evening entertainments: an out-of-doors barbecue to help get better acquainted; a half day off for shopping in Lusaka;[12] a video show of a SWAPO refugee camp; another video of the workshop participants as they debated issues;[13] a half-day trip to an ANC farm which produces food for South African refugees and teaches managerial skills for post-liberation South African agricultural development; and a Saturday night party in Lusaka where the visitors met non-participating Zambians, ate snacks and danced together into the early morning hours. As is often the case, these extra-curricular activities created opportunities for shared experiences and informal exchanges that helped to bridge the differences typical of any group of people who come from several countries.

It took time for the participants to understand each other and the aims of the Learning Process well enough to get on with designing the next steps. Initially, for example, some of the project representatives viewed the Lusaka workshop primarily as their chance to convince the donors they needed more funds. The first day, several argued that their main difficulties stemmed from lack of money. Only after considerable discussion and debate did they realize that the Learning Process might provide them with a more valuable opportunity to talk with and learn from the participants from the other countries. Gradually, they began to work together with others from their own country, including national researchers, to plan the detailed pilot Learning Process. Over time, they realized that, whichever country they came from, they all faced very similar problems, and could benefit a great deal from exchanging ideas about different ways of tackling them. As one said,

> We need more workshops like this. We can learn from our friends from the neighboring countries about what they have tried that works, and what doesn't. When any of us start a new project, we shouldn't have to reinvent the wheel.

By enabling them to meet and talk seriously with villagers about why they had started particular kinds of projects, and the difficulties they encountered, the workshop also gave the three national researchers a different perspective and new ideas. Although the researchers came from primarily rural backgrounds, their years of socialization in the formal educational system[14] had done little to help them acquire the skills needed to work together with peasants to design and carry out the Learning Process. During the workshop, they discovered new ways of thinking about the relationship between national development programs and grassroots activities. Towards the end of the week, one remarked, "This is a different kind of research project than I have ever worked in before."

The donor and intermediary agency staff members began to build new relationships with the local researchers and the project representatives, as well as with each other. To do so, however, they not only had to reject consciously—as most already had done—the role of the donor who controls resources and hence makes the key decisions. They had to root out subconscious attitudes that might perpetuate paternalistic donor-project member relationships. They learned they had to stop using the terminology that permeates the aid community, for words like "client" and "target population" tend to perpetuate dependency relationships.

The workshop participants specified the aims of the Learning Process as:

i. to help private voluntary organizations understand how local, national and regional factors facilitate or impede development projects so they may work with project members to devise more responsive programs;

ii. to empower project members to participate in more systematic analyses of causes of the problems they confront in order to formulate strategies to ensure that aid enables them to achieve increasingly self-reliant development; and

iii. to forge a partnership between national researchers, project members and intermediary agencies, wherever possible using local and regional resources.

To achieve these goals, the workshop agreed that the pilot Learning Process should incorporate the following features:

i. It must be on-going. The Learning Process starts whenever a community realizes it faces a problem which leads to the formulation of a project proposal. The project members, the intermediary agencies, the donors and the researchers should work together at every stage to conduct the process consciously and systematically, revising their initial strategies as needed to solve the new difficulties they inevitably encounter.

ii. Since donors and project members tend to have significantly different perceptions and do not share equal access to resources, the design should involve project members in all aspects of decision-making concerning the Learning Process.

iii. Since projects (including those represented at Lusaka) differ significantly, the learning process should vary with each one. In each project, the project members need to work together with the researchers to design the particular process for their project. In so doing, they will acquire the skills to continue the process after the pilot project ends.

iv. The researcher, acting as a facilitator,[15] should work together with the relevant representatives of the projects and the intermediary agencies to select two persons to live on each project and work with its members for a period of up to two months to develop and implement the Learning Process. In Tanzania and Zambia, the workshop agreed, these would probably be senior university students, carefully selected for their interest and ability. At least one should come from the project area. The second might come from another part of the country to provide an independent perspective. The Zimbabwean representatives, on the other hand did not specify whether to employ university students or local teachers or social workers from the project area. They maintained that the latter might better assist the project members to carry on the Learning Process after completion of the pilot phase.

IMPLEMENTING THE PROCESS

For the rest of 1984 and the first half of 1985, the Lusaka Workshop participants, in cooperation with their co-workers in their home countries, implemented the next stages of the pilot process.

The selection of facilitators

The researchers selected and trained facilitators to live and work on specified projects, participating with the members to design and carry out the proposed problem-solving evaluation. In the event, in all three countries, senior university students played this role. The university student body formed a convenient, large pool of candidates. The researchers could more easily contact, train and supervise them. The students undertook the work during their long vacations. Since the experience contributed to their education, they were willing to work for somewhat lower stipends than would qualified personnel from the community. Perhaps most important, the students, as future development planners and administrators, learned from this hands-on experience the advantages of involving the villagers in analyzing and improving their efforts to raise their own living standards.

In most cases, partly due to financial considerations, the national researchers assigned only one student to each project. Whether this was desirable may be argued. On the one hand, it imposed greater responsibility on each student; on the other, the student had little alternative but to forge close relationships with the project members.

In selecting the students, the researchers employed four criteria:

i. They should be interested in working with rural project members;

ii. they should come from the project area, speak the language, and understand the culture;

iii. they should be social scientists; and,

iv. for women's projects, female candidates should receive priority over male candidates.

In southern Africa, perhaps more than in other Third World regions, many if not most university students come from peasant households. Many tend to be older and more mature than undergraduates in the United States or Europe. As a result, the researchers could relatively easily select qualified students who came from the project areas, spoke the project members' language and understood their culture. For example,

- Before attending the university, the Zambian woman student researcher had worked for almost a decade as an agricultural extension officer. She already had a considerable knowledge of the special difficulties rural women face in working together to increase their incomes.

- Another Zambian student had only managed to get an education because he and his mother scrimped and saved to put him through school; his father wanted him to work on the farm. After the student worked with the women in the uniform sewing project, he said he had come to realize that—contrary to tradition—men could work well with women. He planned to go back to his own village to help his mother to organize her friends.

- One of the Zimbabwean students had left secondary school before independence to join the liberation forces in Mozambique. After working with peasants in the Learning Process, he asked the donor organization if he could obtain a job working with them when he graduated because he wanted to continue to help peasants build grassroots development.

Coming from rural backgrounds like these, the students who took part in the pilot Learning Process could and did empathize with the project members with whom they worked in attempting to discover the causes of the difficulties they faced.

Because of historical factors hindering women from attending the national universities, especially from neglected rural areas, the researchers could not always find qualified female students to work on women's projects. In Zambia, the researcher discovered few qualified women students to work on one of the relatively remote rural women's projects. The Tanzanian researcher identified one qualified Kigoma woman student interested in the Learning Process; he assigned a woman from another region to work with her, but, since she did not speak the local language, the two worked together for a month on each of the two Kigoma projects.

Introductory national workshops

To introduce the national Learning Process in Zambia and Tanzania, the national researchers and intermediaries initially organized three-day national workshops. These served much the same function for each country as the Lusaka workshop did for the regional process. The representatives of the national projects and intermediary agencies met with the students to discuss and learn together about the possible causes of and solutions for the problems plaguing their projects. Together, they sought to determine the best ways of gathering the evidence required to test the most probable explanations.

These national-level workshops enabled the representatives of the local projects to take part in the national learning process from the outset. In Tanzania, the workshops debated issues in Swahili, since the villagers spoke more fluently and freely in their national language. The chairperson they elected, a Kigoma villager, could not speak English.

Although all the Zimbabwe project representatives took part in the Lusaka regional workshop, they had no opportunity to meet the student facilitators in an initial national workshop to discuss and clarify their roles. Instead, the students received informal instructions prior to working with the project members.

Involving the project members

After the introductory national workshops, the project representatives brought the individual students home and introduced them to all the other project members. Where possible, they called a full membership meeting to explain the purposes of the Learning Process and the students' expected role. They helped the students find places to live, whenever possible in the homes of the project members. Although the students paid for their keep, they usually helped their host family with the household chores like any other member of the family. They also took part in the physical work of the project: sewing uniforms, planting and harvesting crops, learning to make hoes. This enabled them to chat informally with the project members about how they viewed the project's activities. They worked with the villagers to gather the relevant evidence to test the suggested explanations of their difficulties. The students kept notes of their discussions with the project members, the kinds of evidence they uncovered, and the implications of the members' differing views of that evidence.

In some cases, because the Zimbabwean project representatives had not met the students at an introductory workshop, it took somewhat longer for them to get settled down and win the members' confidence. Despite these drawbacks, they ultimately worked out fairly successful relationships and played useful roles. Nevertheless, their experience underscored the need for more careful attention to training the students and establishing a good relationship between them and the project members, two objectives successfully achieved by the national workshops in Zambia and Tanzania.

The national researchers' continuing role

The national researchers supervised the students. Where possible, they visited them at least once during their stay on the projects.[16] Halfway through the 6-8 week period during which the students lived and worked on the projects, the national researchers brought them together with the relevant intermediary staff to analyze their progress and to share ideas as to more effective ways to involve the project members in the process.

Discussing the findings with the project members

At the end of the 6-8 week period, in those projects involving relatively few people, the project leaders called the members together to discuss the findings of the investigative process. One of the members, the students, or both, reported on the tentative findings and suggested strategies for dealing with the difficulties raised. Those present at the meeting criticized and made suggestions for improving the final report.

The final national workshops

In Zambia and Tanzania, after completion of the students' work with the project members, the researchers and intermediary agency representatives arranged a second national workshop. This enabled project representatives and the students to meet the national researcher and the intermediary staff members to discuss and compare their overall findings. They examined the evidence for systemic causes of problems to lay a foundation for proposing more self-reliant strategies. They also reviewed the entire national pilot process to consider whether and how to institutionalize it in each country. In Tanzania, the second workshop took place in the central province to enable the participants to visit the food-aid project and discuss the findings directly with project members.

In the discussions at these final national workshops, the national researchers brought to bear a broader range of evidence and theoretical background in examining the underlying nature and systemic causes of the problems which seemed to plague all the projects. At the same time, they themselves acquired more information about how local, national and international policies and programs affected rural life. The intermediary staff members presented useful suggestions concerning available resources and experiences of similar projects elsewhere. They too gained greater insights into how the projects actually functioned.

The on-going Learning Process

On their return from the workshops, the project representatives often found themselves able to work more effectively with other members to define more clearly their projects' goals and work out better ways of achieving them. A Kigoma regional official later reported that the Learning Process experience had enabled the Kigoma representative, who chaired the Tanzanian workshop, to explain more clearly to the district authorities the villagers' needs and concerns.

In Zimbabwe, instead of a second national workshop, the students met with the national researcher and the intermediary staff members to report on their findings. This meeting did not include the project representatives, who only met again as a group for one evening at the final regional workshop. Comparison of the procedures in the three countries suggests that the inclusion of project members in follow-up national workshops increased their benefits from the process. The in-depth give-and-take with participants from their own country at the final national workshops enabled them to explore in more depth how to analyze and deal with their own difficulties.

THE FINAL REGIONAL WORKSHOP AT GWEBI

At the end of the year, in a final regional workshop held in Gwebi college in Zimbabwe, representatives from all the projects included in the pilot process met with representatives of the intermediary agencies and the national researchers to review the findings from all three countries. This time, most of the workshop participants already knew each other and eagerly shared their previous year's experiences. The first day, they again elected a steering committee which chose as overall chairperson a woman intermediary representative. The steering committee again drafted an agenda which the participants adopted the following day.

As in Lusaka, the workshop steering committee arranged various free-time activities to give the participants time to chat informally and plumb more deeply into issues that concerned them. These included a tour of the premises of the Gwebi agricultural college;[17] a movie showing grassroots aid projects; a half day shopping trip in Harare; a Saturday evening barbecue where members of non-government organizations' Harare staff, interested in the Learning Process, met and chatted with the workshop participants; and a day-long trip to visit and learn more about the cooperative tractor project in northeastern Zimbabwe.

During the first half of the week, the workshop participants critically reviewed the year-long Learning Process. They recognized that the Lusaka workshop did not create a "perfect" design. Like any other development strategy, this one too inevitably encountered difficulties. The participants of the second regional workshop identified some of these, and proposed new ways to deal with them. Others remained for participants in a future learning process to solve.

The participants in this second regional workshop concluded that the southern African Pilot Project had been useful in strengthening their national and regional capacities to implement and evaluate self-reliant grassroots development. In particular, it showed the advantages of involving the project members together with national researchers and intermediaries in evaluating the impact of aid on development. They formulated a proposal to bring together more national intermediaries and research institutions to extend and improve the participatory, problem-solving approach they had initiated.

The next chapter describes the main findings of the year-long Southern African Pilot Learning Process. The final chapter outlines the main recommendations of the Gwebi Workshop participants as to how that process might be improved in the future.

CHAPTER FOUR
Notes

[1]This term should be distinguished from the concept as related to urban lending; here it refers to the kind of participatory problem-solving evaluation of experience described in Chapter 2 above.

[2]As part of the preparation of the learning process, the Oxfam America staff organized the Applied Development Research Network to bring together interested people in the United States who had conducted research, taught, or worked in the development aid community in Africa. Over time, the network members, through such activities as the Learning Process, hoped to establish collegial relations with researchers and development workers in Africa, thus improving the background information available to U.S. private voluntary organizations.

[3]See appendix for list of members.

[4]Zambia, August-September, with two additional weeks during Christmas break; Zimbabwe, January-February; Tanzania, April-May.

[5]Several private voluntary aid organizations, including Oxfam America, provide funds to the ANC and the Southwest African People's Organization (SWAPO) for refugee projects. Both sent representatives to the first workshop, but unfortunately, primarily due to increased South African destabilization tactics (for details see A. Seidman, *The Roots of Crisis in Southern Africa,* Trenton: Africa World Press for Oxfam America, 1985), they could not devote the time or resources over the course of the next year to implementation of the learning process.

[6]See asterisks by projects listed in Table 4.1

[7]Neither representatives of this program nor the researchers who took part in analyzing it had participated in the Lusaka workshop. Since no national workshop took place in Zimbabwe, they had no opportunity to share with the others in the formulation of the research process. As a result, although two students lived in the communal area for many weeks, they did not work with the members of a few selected projects to achieve an in-depth, problem-solving analysis. Instead, they attempted to survey 64 projects. While it seems doubtful that this process contributed much to raising the consciousness levels of the project members, nevertheless the conclusions reached tended to coincide with those found by the other project participants.

[8]In theory, the ten house cell system, organized by the ruling party, operates throughout the rural areas in Tanzania. Roughly every ten households form a cell which elects a leader to represent them in village decision-making bodies.

[9]PL480 refers to the United States law which provides for the distribution of surplus U.S. food crops to countries in need. Some, as in this instance, is distributed through private voluntary organizations.

[10]See Table 4-1.

[11]As he did on several occasions, translated by the national researcher or the intermediary representative from Zimbabwe.

[12]The regional workshops provided several participants their first opportunity to visit a neighboring southern African country, and many wished to bring mementos to their families.

[13]An ANC camera team made video tapes of several of the workshop sessions and later took pictures of one of the Zambia projects. These tapes are available for future use as training materials.

[14]Two of the three had spent years abroad getting their PhDs; the third had completed his Masters in the national university.

[15]That is, helping to organize the Learning Process, getting it underway through discussions with relevant people leading to appropriate next steps, and initiating interactions among the different participants.

[16]Some of the projects were in such remote areas that this proved more difficult than anticipated, especially because lack of fuel and spare parts made transport difficult.

[17]Until independence, the college had admitted only white males as agricultural extension trainees, primarily to service the country's 6000 white commercial farms; now, with a new black principal, it trains Africans to work with the more than 800,000 peasant familes as well.

Cooperative members headload buckets of water to irrigate their plots. (Photo by Michael Scott, Oxfam America)

CHAPTER FIVE
The Findings

We discovered some of us had already found better ways of dealing with the problems we have. By talking over our findings together, we all learned how we could do a better job.

—A project representative at the Gwebi workshop.

INTRODUCTION

The Southern African Pilot Learning Process Project provided a test of the propositions outlined in Chapter 2 as to the advantages of a participatory, problem-solving process to assess aid's impact on grassroots projects. Those propositions hold that:

1. Only through the participation of the project members can the process contribute to an adequate evaluation of all the factors influencing a project's success or failure;

2. A participatory evaluation requires a problem-solving methodology, as opposed to an ends-means approach;

3. An outside facilitator is helpful in assisting project members to acquire the skills necessary to participate effectively in the evaluation process.

The results of the pilot Learning Process substantiated these propositions. On the one hand, the findings revealed systemic internal problems that an outsider, alone, could not discover. The projects will likely fail unless their members understand and learn to deal with these kinds of problems. On the other hand, working together with national researchers, project members can improve their own capacity to analyze the causes of the problems they confront and design better strategies for overcoming them.

The pilot project dealt with 14 quite different kinds of projects in 3 countries. Limited space prohibits presentation of the detailed findings.

Nevertheless, a summary serves to illustrate the potential of this kind of participatory evaluation.

Underdevelopment implies that, among other difficulties, most rural inhabitants may not have sufficient skills and resources to improve their living and working conditions. Not surprisingly, the pilot process showed that lack of technical and managerial skills and complementary resources hindered the project members from making ideal use of aid. To recount these difficulties for each individual project would not contribute much to either the private voluntary organizations' or the project members' knowledge. Instead, this report focuses on the causes of tendencies leading to the problems plaguing the projects; and the strategies the pilot Learning Process generated to overcome them.

The evidence gathered by the project members and the students in all the projects evaluated in the three countries suggests six systemic tendencies potentially leading to similar problems. This chapter presents the findings concerning these six tendencies:

- The peripheral impact of some aid projects on surrounding communities;
- The formation of elites;
- Conflicts with local government officials;
- Donor policies which foster dependence;
- National and international development policies that leave rural communities vulnerable to external forces, causing chronic difficulties for small rural projects seeking to obtain essential inputs and market outputs;
- The special difficulties confronting women and their projects.

In each case, the chapter first outlines the tendency toward the difficulty as manifested in the three countries. It then analyzes the causes identified by the Learning Process participants. Finally, it summarizes the participants' conclusions as to possible strategies for avoiding the tendency. To facilitate keeping track of the examples used to illustrate each tendency, the reader may wish to refer to Table 4-1 on page 68, above.

TENDENCY #1: A peripheral impact

The seemingly peripheral impact of some projects in terms of their contribution to overcoming the difficulties confronting the poor majority in the surrounding community.

Most aid agencies anticipate that their transfer of resources to particular projects will improve living and working conditions of the poor in rural areas. The small size and limited objectives of some individual projects, however, tends to render their role peripheral to the needs of the community.

For example, in Zambia a group of church women on the Copper Belt had begun growing cotton to raise money for food as a charitable contribution to malnourished children. Later, they shifted to growing vegetables, partly to raise incomes for themselves, though they continued to give food to a neighboring hospital. Prior to taking part in the pilot Learning Process, they had not reached agreement, even among themselves, as to their long-term goals. The two dozen women in the uniform sewing project had developed a fairly successful project to sell uniforms, but this had not yet stimulated other women in the region to undertake income-generating projects. In the northern province, the wives of some employees, together with women workers on a parastatal coffee plantation, proposed to engage in sewing activity to raise their incomes. They had not yet decided what to produce or where to sell what they produced.

The Tanzania food aid project had broader aims, but its implementation did not always achieve them. The women who purchased the powdered milk at subsidized rates occasionally sold it in the market to obtain cash for goods like school uniforms and local foods. They apparently preferred fresh milk to the powder. They did not fully understand or know how to measure the quantities of powder to mix with water, and sometimes used it as a porridge. Nor did they always understand that, to avoid the danger of water-borne diseases, they must boil the water before giving the powdered milk mixture to infants.

The Tanzanian carpentry and timber projects did not receive widespread community support. At first the tools provided lay unused, as the villages had made no arrangements for training their users. Eventually, the village leaders loaned the carpentry tools to self-employed carpenters, until they realized that these men were using them to replace their own worn-out equipment. The village leaders then tried to mobilize un- and underemployed primary school dropouts to undertake training. The first year they failed to attract any; the second year, they persuaded about ten to undergo training and gave them the tools, but five dropped out. The parents did not encourage boys to take part. They needed their labor during the farming season. If the boys were to undertake training, their parents preferred them to enter more prestigious programs with higher potential returns.

In the timber case, the village leaders provided the tools to a number of loggers, some of whom had worked for a neighboring mission to set up business on their own. The men worked at logging during the season when their wives did most of the farm work. However, this neither trained new workers, nor provided new employment. Furthermore, the district authorities refused to contract with the timber workers to cut timber for district bridges, a substantial market. Instead, the authorities gave the bridge contract to private construction companies who skimmed off the profits by purchasing half the timber from the project members.

Although designed to provide food supplies and income for entire villages, the Kigoma maize-beans projects likewise lacked community support. The village women, who do a substantial share of the planting, weeding and harvesting of food crops, said they knew little about the projects' aims. All the villagers objected that the cultivation of maize and beans competed for time they needed to work in the family fields. In addition, the district leaders required them to take part in another community program to grow cotton for export. Some asked if they could not substitute the beans-maize project for the required cotton growing; at least, they said, if they could not sell the maize and beans, they could eat them.

In Zimbabwe, the projects examined more typically sought to encourage villagers to branch out into new kinds of productive and development-oriented activities. From the outset, these projects seemed designed to involve the larger community. The tractor scheme aimed to involve some 200 families in 10 villages in the purchase of tractors for use in plowing. The revolving credit scheme aimed to provide funds for a number of women's groups to finance their own projects. The training institute was started as a result of villagers' discussions of the need to provide skills to members of the younger generation who could then return to their villages to help implement development projects. As part of a program conceived during the liberation war to end wealthy farm owners' employment of farm workers at below-poverty-line wages, several ex-freedom fighters agreed to form a producer cooperative to help maintain and increase productivity. A Zimbabwean, whose community had been disrupted during the liberation war, initially requested donor assistance to undertake projects to enable the community to recover; over time, he and his co-workers evolved their initial efforts into an integrated development program that spread throughout the communal area.

A Tentative Explanation for Tendency #1

Donor organizations and project holders sometimes develop proposals for assistance without adequate involvement of the community members in thinking through criteria and strategies related to their particular needs and the difficulties they confront. Therefore, the resulting projects may have little multiplier effect in generating employment, incomes and improved living standards. In contrast, where rural communities have participated in a prolonged self-reliant struggle for their own liberation, as in Zimbabwe, their consciousness of the potential of community-wide efforts may foster their own initiative in undertaking broader development efforts.

To illustrate: The selection of the women's projects in Zambia seemed often to have been made as a result of acquaintance with the leaders of existing women's groups. This may be a necessary first step, but does not provide a channel for involving the majority of rural women in analyzing the kinds of activities in which they could engage to improve their lives.

In several Tanzanian cases, the donor organizations discussed the projects they would fund with a national intermediary agency. The intermediary worked out details of the projects with district leaders, and through them with village leaders. Although the village leaders generally maintained they had discussed the project with the villagers, village members usually had no knowledge of the projects' aims. Some expressed little interest in the projects' outcomes. They had seldom participated in planning the projects. As a result, they may have worked only half-heartedly to implement them.

For example, the idea of the beans-maize project seemed to have been initiated during an earlier World Bank program in the region.[1] The villagers had requested assistance to grow maize, but when the aid came through, the beans component had been added. The village leaders indicated that, had they been adequately consulted, they would have argued for a tractor rather than oxen because of the danger of the tsetse fly; they maintained the death of six of the oxen in one of the villages proved their point.[2] On the other hand, the village leaders, all men, had failed to involve the women, who do a substantial share of the farm work, in planning the program. As a result, the villagers did not fully understand the project's objectives and did not always perform the required tasks.

Suggested Strategies to Overcome Tendency #1

Donor organizations should involve potential project holders, perhaps with the assistance of national researchers, in a detailed, participatory assessment of the constraints and resources characteristic of their particular community; and conduct adequate research concerning the sources of supplies and available markets for proposed income-

generating projects. On this basis, community members should design their own projects to meet their basic needs as they see them.

The Zimbabwe projects' experiences illustrate how this might work. Community leaders, many of whom had participated in the liberation struggle, typically worked together with community members to formulate the project proposals. In the revolving credit scheme, elected representatives of the women decided which groups to lend funds to. They started initially with vegetable gardens, but the creation of too many gardens led to marketing problems. After analyzing these difficulties, several groups branched out to establish open-air bakeries and chicken-raising projects. The availability of the revolving credit fund gave them the flexibility to multiply their own efforts and involve other women in the community. Trained women community workers (see below re: strategies to overcome tendency #2) helped to ensure the proposed projects' viability. Seeing their success, a number of unemployed men in the communal area, who had originally held aloof from these "women's" projects, asked permission to join them.

The leaders of the Zimbabwe umbrella organization always meet with relevant group members before seeking aid to undertake a new project like the training institute. They discuss how the project will work, each group member's role in it, and anticipate difficulties as well as rewards. This way, the groups themselves decided that the institute should train young women and men to make and lay bricks for community projects; to make furniture and simple farm implements; to sew school uniforms and other items for community members; and eventually to weld and repair more complicated farm machinery.

Several ex-combatants brought their relatives and neighbors together to form the Zimbabwe producer cooperative. Their liberation war experience had convinced them that only in that way could peasant farmers achieve self-reliance and improved living conditions. Through cooperatively pooling land and income to buy farm machinery and other modern farm supplies, they could compete successfully with the white-owned commercial farms that still dominated the national agricultural scene.

The Zimbabwe tractor scheme emerged from the peasants' need for draft power to replace the cattle they had lost during the war. The head of a nearby state farm, who had taken part in the liberation struggle, helped them form 10 groups of 20 families each and approach a private voluntary agency for funds to buy a tractor for each group. The agency provided the down-payments, and assisted the peasants in negotiating for loans with the government-owned Agricultural Finance Corporation. By cooperatively cultivating 20 acres and selling the produce, each group planned to earn funds to pay off the loan. In addition, the groups then rented the tractors to members (and to non-members, at a somewhat higher rate) to plow their own land.

The impetus for the spread of the integrated communal area program came from inside the community itself. All the participating projects elected representatives to a village committee that encouraged villagers to visit existing projects and to discuss their own ideas for ways to generate incomes. The village committee assisted in starting new projects, helping them to design the necessary inputs and discover markets, as well as to establish democratic and participatory methods of working together. After the new project members had shown their determination to work together with their own resources, the village committee leaders would help them to find additional aid if necessary.

As a strategy to broaden their projects' impact, the second Tanzanian workshop proposed to link similar projects. For example, the carpentry and the timber projects, located in neighboring southwestern villages, could work together. The carpentry project might purchase timber; and in turn the carpenters might produce furniture, windows and doorframes needed to improve the timber workers' housing.

At the final regional workshop, participants pointed out that the introduction of a participatory learning process in the initial stages would help improve the selection and planning of projects, and stimulate a spread effect. The participants learned that, in Dominica, when two or three people asked the Small Projects Assessment Team for assistance to start a project, the team first visited the site and discussed it with all the potential members. Once convinced of their grassroots base, the team members helped them design the project and obtain funds. In India, donor organizations sometimes fund workshops for representatives of similar types of projects to exchange ideas and consider alternative strategies.

Following their analyses of these experiences, the Learning Process participants proposed broadening the scope and methodology used in making feasibility studies: Would-be project members should work together with national researchers in a systematic analysis of the constraints their proposed project might confront, and how they might use their existing resources to overcome them. They could consider the accumulated experiences of similar projects elsewhere in the region, as well as relevant information available from national and regional research institutions. On this basis, they should decide what additional resources they might need, and the specific role aid might play. In the process, the project members themselves could think through how their projects could contribute more effectively to self-reliant development for themselves and their neighbors.

TENDENCY #2: The formation of elites

The formation of elites among the leaders of the projects, combined with the danger of corruption and possible misuse of resources.

In most projects, the leaders tended to acquire elite status. They made most of the decisions and controlled the projects' resources. Meanwhile, at least some members voiced suspicions that the leaders were manipulating the projects for their own ends.

For example, the managers of the hoe-furniture industry, who often received their pay late, frequently took money from the bookkeeper, without adequately accounting for it, apparently for their own or their family members' use. This hindered adequate accounting. It caused conflicts among the managers and suspicions among the workers. It also obscured the fact that the managers consumed part of the project's profits which should have been invested to make it viable. The workers complained that the managers set their own salaries at K490 a month, while paying the workers only K40 to K60 a month, that the managers did not participate in the actual work of the project, and that they took too much of the project's income for themselves. However, the workers feared to speak openly for fear they would lose their jobs.[3]

A similar tendency emerged in a less obvious form in some of the smaller Zambian women's projects. In the vegetable project, the chairperson was the wife of the farmer who provided the land. The farmer permitted the women to use his pump to irrigate their field as long as they purchased the diesel fuel. The women members suspected that he used their fuel to irrigate his fields. However, they did not dare complain openly or criticize what they viewed as their chairperson's failure to protect their interests for fear they would lose access to the land.

The women on the coffee plantation contributed funds to purchase materials for their sewing project. The project initiator, who owned a car, purchased the sewing materials for them but failed to give them the receipts. The women thought she was misusing their funds and thought their chairperson should have insisted on obtaining the receipts. But they, too, feared to complain because the chairperson's husband managed the coffee plantation which employed some of the members and the husbands of most of them.

In two Zimbabwe projects, this tendency appeared in a different form. In the case of the Zimbabwe producer cooperative, misunderstandings arose among the members as to how this type of cooperative should function. From the beginning, the several ex-combatants who organized it sought to maintain collective decision-making and working methods. They conducted classes in the evenings to educate members as to the purposes of collective production and the cooperative's relationship to the larger society. When the Ministry of Lands and Resettlement permitted them to resettle on a larger former commercial farm, they recruited six new members through the Ministry of Labour's unemployment registry. These people, coming from another part of the country, shared none of the original members' knowledge about or zeal for cooperative efforts. They reportedly felt like outsiders and began to resent the ex-combatant leaders' roles. They believed the leaders, who lived in the former owner's house rather than huts like the other members, used the project's lorry not only for project business, but for their own profit.[4] Their resentment flared when the drought and heavy interest payments on Agricultural Finance Corporation loans made it impossible for the cooperative to pay them increased cash incomes.

In the Zimbabwe tractor scheme, the sale of the cooperatively cultivated crops proved inadequate to cover the operating costs of the tractors. These could not be paid for with the short-term credit available to members. The tractor plowed two acres for each member family free. After that, all members had to pay cash to use the tractor to plow additional land. The poorer members could only hire the tractor to plow a few acres, sometimes too late in the season to be of any use. Some members complained that wealthier members, or those with larger families, sent children or even hired labor to work on the cooperative plot, while they had to contribute their own labor. Poorer members resented this work even more since the wealthier members could afford to hire the tractor to plow more acres of their own land.

In Tanzania, the villagers reportedly did not voice similar suspicions of their leaders. However, the village leaders in one of the Kigoma villages used the profits from the sale of the maize the villagers grew to help finance construction of an office building for Chama cha Mapinduzi (CCM),[5] instead of financing cultivation of the land and fertilizer and seed for the next season. This may have contributed to the villagers' reduced willingness to grow maize the next year.

In one of the two Tanzanian food aid projects, the donor agency—a religious order—arranged for the storage and sale of the surplus food with the village-level religious leaders. It is not clear that the village leaders played any role in the initial planning of these projects. The women villagers reportedly suspected some village leaders had sold some of the food for their own profits, and misused the funds accumulated from the women's own purchases.

The Tentative Explanation for Tendency #2

> *Learning to work together as equals in any kind of project as required to make the best use of aid, takes time. Peasants, like most other people, have typically been brought up in hierarchical societies. The very nature of underdevelopment implies that the majority of project members lack necessary skills and access to resources. A project's leaders, who typically have attained their status by virtue of the fact that they have more skills or access to resources than others, may emerge as an elite in control over the project's decision-making. The leaders may seek to benefit themselves and even engage in corruption. On the other hand, the rank and file project holders, who have less resources or management skills and may not fully understand the complex aspects of the projects, or may be excluded from decision-making, may suspect their leaders of corruption—and occasionally their fears find justification in fact.*

To illustrate: The entrepreneurs who set up the hoe-furniture industry in eastern Zambia had already acquired the necessary skills and brought with them their own tools and equipment. One had also acquired a license which, in Zambia, is necessary to engage in production for sale. The entrepreneurs became the managers of the small industry. The one manager who spoke English conducted most of the negotiations with the donors, from whom he obtained capital equipment as well as uniforms for the workers. Thus the donor contributions apparently replaced the need to save and invest the project's own profits. The manager, too, improved his relations with the workers by obtaining uniforms for them. This gave him a bargaining advantage in conflicts with his partner.

The women in the vegetable-raising project were afraid to complain openly about their chairperson's failure to protect their interests, since her husband owned the land they cultivated. Similarly, the Kasama women did not dare complain openly of their suspicions about their chairperson's misuse of their funds, since she was the wife of the manager of the coffee plantation. Many of the members either worked on the plantation themselves or were married to men who did.

The tractor scheme's working rules fostered a tendency towards internal conflict, leading some members to resign. The cooperatively farmed plot's output only covered the cost of repaying the initial loan; to pay for fuel, spare parts and repairs, the scheme charged a fixed fee per acre for plowing private land beyond the first two acres. This fee had to be paid in cash in order to finance the day-to-day costs of running the tractors. As a result, only those members with ready cash—local storekeepers, more well-to-do farmers with savings, or families with relatives earning fairly high wages in the "modern" sector—could afford to hire the tractor. The more cash they had, the larger the acreage they could plow and the larger the harvest they could sell. Poorer members complained that, although they farmed the cooperative plot to finance the tractor loan, they could not benefit as much as more wealthy members from its use.

The tractor scheme tended to foster stratification in the larger community. Even non-members who could afford to pay cash could hire the tractors, although at higher rates. The wealthier non-members who hired the tractors to clear and plant larger acreages, used their profits to finance rebuilding their cattle herds. This provided them with an alternative source of draft power. Poorer non-members, on the other hand, would either have to rent cattle from them, perhaps paying for their use with their own labor, or cultivate only with hand tools.

In the Tanzania food-for-development projects, the women viewed the projects as controlled by the village leaders. The donor agency perceived women and their children as the main beneficiaries of the projects. Nevertheless, the village leaders did not involve the women in decision-making concerning either the kinds of food made available through the project, or the purposes for which the proceeds from the sales were made. The women believed that the goats, purchased with the project's income, belonged to the village, rather than to themselves.

This tendency appeared less pronounced in the Tanzanian *ujamaa* villages where most of the resources (beans and maize seed, oxen, tools, carpentry and logging tools) were received in kind through the district councils. This reduced opportunities to misuse the funds. Furthermore, the 10 house cell structure may have mediated internal conflicts which might otherwise have arisen. The village leaders elected through that structure seemed to possess a kind of legitimacy: in a sense, the villagers viewed the leaders as acting on their behalf.[6] However, the effectiveness of this structure depends on the extent to which it truly incorporates all the villagers' concerns. The village members did not complain or express suspicions about their leaders' role; however, they did indicate lack of information about the project and objections to the additional work-load. This seemed to reflect their lack of participation in planning the project. (See tendency #6 below for the extent to which women tended to be excluded from decision-making).

In all these cases, the tendency for project members to suspect their leaders emerged where the project leaders had not involved them adequately in decision-making. The members depended on their leaders for access to resources (often including the aid the donors provided), and lacked the skills—especially bookkeeping—required to monitor them.

Suggested Strategy for Overcoming Tendency #2

> *By participating over time in a systematic investigation of the causes of their project's internal conflicts, the members can learn how to avoid them. They may formulate working rules that prevent their leaders from acquiring elite status and their project from generating stratification. They can ensure that all members learn managerial skills, including "economic literacy" and basic bookkeeping techniques. This will enable them initially to monitor and eventually to play a more direct role in governing their project.*

The above explanation suggests the necessity for project members, together with the donor agency (possibly with the assistance of national researchers), to design and institutionalize methods to ensure that the project's working rules counter tendencies towards elite formation and stratification. From the outset, they need to involve all the members as well as the leaders in discussing how the members will participate and benefit from the project. Where necessary, the initial plans should include training an increasing number of members in the skills necessary to enable them to take part in making decisions and monitoring the project.

In three out of the four Zambian cases, for example, the students helped the project members to analyze the danger of the tendency towards elite formation. Their reports also alerted the relevant donor agencies and their intermediaries. Participatory discussions involving all the project members with the intermediaries led to formalization of new working rules and procedures that helped to ameliorate the situation. However, in the northeastern project, where the women had as yet received no aid, no particular donor or intermediary took responsibility for working with the group. The group's internal conflicts ultimately led most of the members to drop out.

The final regional workshop debated the case of the small-scale industry at further length. Some argued that the enterprise should function as a cooperative. Others suggested that whether a small industry functions as a cooperative or a modified private firm depends on the stage of social transformation in the country: In some stages, small private industries may provide the stimulus necessary for rural development. Only at a later stage would the members have the necessary skills to run it as a cooperative. This reemphasized the value of careful participatory investigation of the socio-economic conditions within which an industry operates.

In the case of the Zimbabwe women's credit fund, the Ministry of Women and Community Development trained women community workers chosen by the women themselves. These lived in the community. Each was assigned to maintain close contact with 15 to 20 groups of women, helping them to work out the details of their proposed projects. If members had suspicions about their leadership, (which in at least one reported instance they did) they requested the relevant community worker to help them to deal with the situation. In addition, the credit scheme provided funding for training the women to monitor their projects' use of resources. Among other aspects of their work, the Zimbabwean non-governmental umbrella organization that operated the training institute underscored the importance of democratic participation by all its members in all decisions.

In the case of the Zimbabwe tractor project, the student proposed several alternative strategies for the donor agency and the project holders to consider for altering the working rules to prevent stratification. When the final regional workshop participants visited the project, they emphasized the need for all the project members, together with the donor, to reexamine the scheme's working rules to ensure that it contributed to the poorer peasants' self-reliance and higher living standards. In particular, they suggested expansion of the revolving fund[7] to provide credit to enable project members to pay for greater use of the tractor.

These findings suggest the benefits of involving all project members in planning and implementing projects so that they fully understand the project's goals and what they must do to help achieve them. At the same time, plans for the projects must include training programs to ensure that all project members acquire skills in order to eventually participate fully in controlling and developing the resources to which the projects give them access.

TENDENCY #3: Donor policies which foster dependence

The emergence of conflicts over resources between projects receiving aid and local government agencies.

Project members complained that local authorities sometimes refused to assist them by providing resources under their control, such as buildings and transport. Some expressed fears that the local government agencies might take advantage of their access to outside aid. On the other hand, local government officials sometimes complained that project members did not adequately consider the projects' perspectives in terms of government development plans.

Incipient forms of this tendency appeared in all four of the Zambian projects. The managers of the hoe-furniture industry noted that a local member of parliament had helped them make contact with donor agencies. The district officials showed the industry off to visitors as an example of development in the province, but otherwise gave them no practical support. Members of the uniform sewing group argued that, since they were performing a community service, local authorities should let them occupy a vacant building rent free.

In Zimbabwe, the head of a nearby state farm aided the tractor cooperative in finding a donor agency to help get the project started. But he and the donor agency had to negotiate for nine months to convince the officials of the Agricultural Finance Corporation (AFC)—established by the previous regime to provide credit for white commercial farmers—to lend funds to the peasant scheme in a communal area. The peasants found themselves saddled with debt repayment, including a high rate of interest which consumed most of the surplus they earned by selling the produce of their cooperatively cultivated plot. This left them with no funds to purchase diesel fuel, spares, etc.

The western Zimbabwe umbrella organization's training institute encountered semi-official governmental opposition which reflected the ethnic-political conflict that disrupted the region after independence. The organization had located the training institute in a communal area into which the colonial-settler regime had pushed both Ndebele and Shona peasants. The non-governmental organization set up the training institute in the Ndebele community. Local government officials began to express suspicions that the training institute was a ZAPU front and threatened to close it down.

The ex-freedom fighters' producer cooperative also encountered difficulties with the state authorities when six members of the cooperative complained to a government official. With very little if any investigation, he accepted their allegations, suspended the cooperative leadership committee and appointed an administrator.[8] The administrator had no apparent interest in the cooperative. He appointed the head of the suspended committee as his assistant to supervise members' work, but that satisfied no one. Work discipline deteriorated. Finally, the original members of the cooperative gathered their belongings and prepared to return to their home area, hoping to acquire their own land free of government intervention. The few remaining cooperators agreed to let them take some fertilizer as a partial compensation for the loss of their demobilization pay and several years of hard work in building the cooperative. The government official, however, refused to let them remove it.

In Tanzania, where the inputs to village projects usually came through district councils, the villagers tended to blame the district officials for the difficulties they encountered in their projects. The Kigoma villagers complained that district officials did not send the seed and oxen on time for the agricultural season. The village leaders in Songea claimed that district officials refused to permit the timber project bid to supply timber for government bridges; but the private contractor who won the bid later purchased half the timber from the timber project.

In contrast, the donor agency administering the food aid project in central Tanzania worked directly through village-level church leaders. In that case, little conflict occurred between the project leaders and the district officials.

A Tentative Explanation for Tendency #3

> *By virtue of their control of resources as well as their relatively high incomes, the officials running local state agencies may constitute an elite strata in the typical rural community. This may have contradictory implications for projects receiving outside aid. On the one hand, such officials may support projects which provide an appearance of development in line with stated national objectives, but they may give first call on the resources they control to their own programs, or, in some cases, to well-to-do or politically powerful elements. On the other hand, they may perceive projects which receive resources from outside aid agencies as a threat to their own status, since such projects may establish a power base outside their control. Ideological or ethnic differences may aggravate the consequences of these tendencies.*

In Zambia, for example, the women's vegetable-raising group discovered that they could not obtain tractors or bulldozers in time for the planting season because the government land-clearing unit had already allocated its equipment to three neighboring commercial farmers, all of whom happened to be high-level government or party officials.

In Zimbabwe, on the one hand, the unusually good relationship between the women's revolving credit fund and local government officials apparently reflected the fact that during the liberation war their communal area had strongly supported the Zimbabwe African National Union, which subsequently became the ruling party. On the other hand, the local authorities' distrust of the training institute in midwestern Zimbabwe apparently reflected the unrest and destabilization in that region. In this atmosphere, up to 100 South African agents (according to Zimbabwe government estimates) intervened to further destabilize the area, attacking development projects, killing villagers and white commercial farmers.[9] The government repeatedly imposed curfews. The training institute appeared likely to become a victim of the resulting tense atmosphere.

The Zimbabwe producer cooperative's experience illustrates another feature of the contradictory post-independence state in Zimbabwe. Although he was black, the local government official had recently purchased a commercial farm in an adjacent area. He had already objected that the cooperative was not functioning as required by the existing cooperative law. Passed by the pre-independence regime, that law—contrary to the new government declared policy—limited cooperative functions to service, rather than productive, activities. The local official may have feared the consequences of a successful producer cooperative and its leaders' efforts to organize farm workers in the region. Instead of assisting the cooperative's members to overcome the inherent tendency towards internal conflict, he intervened in a manner that led to the cooperative's breakup.

In Tanzania, the relationship of local authorities to aid projects reflected that country's quite different historical experience. Private voluntary organizations that operated directly with the villagers through their own church-related networks experienced internal conflicts between villagers and some village leaders, but none with district officials. In contrast, the intermediary aid agency generally channelled aid through the district authorities, who in turn made goods available to the villagers in kind rather than cash. This policy reflected the agency's lack of sufficient staff to maintain direct contact with distant villages, and concern lest the villagers might use cash for purposes other than those intended. Given an overall shortage of funds and many other essential inputs, villagers complained that these officials followed their own agendas in spending agency funds.

Suggested Strategies for Dealing with Tendency #3

Project members, donor agencies and intermediaries, with the aid of national researchers, need to understand the contradictory features of government agencies, which differ between and within countries. On this basis, they should work together with project members to design appropriate strategies to protect the project from outside domination. Where possible, project members should try to gain access to state provided resources, and to acquire skills to enable them successfully to utilize those resources as well as others transferred from private agencies.

At the final regional workshop, a participant suggested that, in one sense, local citizens spoiled their local officials. They elected them to office and then failed to press them to contribute to fulfillment of community needs. The workshop agreed that project members should maintain consistent contact with local elected officials to ensure that they implement policies that contribute to community development.

In Zambia, after taking part in the pilot Learning Process, the project members, together with intermediary staff, worked to clarify the issues and evolve a more cooperative relationship with local authorities. The women's project officer, for example, worked with the uniform sewing group to obtain materials and transport their products in district authority vehicles when they traveled for other purposes. The women's vegetable-growing group, together with the student, visited a nearby army unit to negotiate for needed draft power. The workshop suggested that the hoe-furniture industry, as a small business, should pay rent for premises provided by the local authorities.

In Zimbabwe, the close liason with local and ministerial personnel facilitated the work of the revolving credit fund in helping women's groups achieve self-reliant projects. In other cases, outside personnel sometimes also proved helpful to project holders. For example, the donor agency together with the management of the neighboring state farm succeeded after protracted negotiations in getting the state-owned Agricultural Finance Corporation to lend funds to the tractor scheme, the first time they had made loans to communal area peasant groups.

In the midwest, when local authorities began to call for closure of the Zimbabwe training institute, the student, who lived with his family in the neighboring community, helped relieve the situation. He had won the confidence of the training institute members, and had become convinced of the institute's potential value to the whole area. He talked with his neighbor, the police commissioner. Together they persuaded the district council to meet to discuss the training institute's potential contribution to district development. The police commissioner made a strong pitch in its support. At least for a time, the authorities agreed to permit the institute to continue to function.

The village leaders of the integrated communal development program consciously cultivated their relationships with local government authorities. The community elected one of them to the local government council. This helped to ensure government cooperation in obtaining locally-available resources.

The pilot Learning Process supported the Tanzanian intermediary agency's decision to decentralize its personnel throughout the country. This would enable them to work in a more participatory mode with the villagers. The second Tanzanian workshop commended this decision, but cautioned that the intermediary should give the local extension officers scope to make essential decisions. Continued control over their activities from the center would cost more in the form of expenditures for phone calls and extra travel to and from the center. The failure to meet project members' expectations, arising from their closer contact with the extension officers, might also undermine their morale.

The second Tanzanian workshop recommended that project members and intermediary staff members also should meet together with relevant government officials in preliminary workshops as part of an initial feasibility study. They should meet again at regular intervals after the project is underway to ensure that all the parties participate in an on-going self-evaluation. At the final regional workshop, however, some participants introduced a strong cautionary note. In countries like India and Dominica, project members often discovered the need to steer clear of local and national authorities. The state, when structured to benefit a politically and economically dominant class, might interfere to block a project's efforts to strengthen the capacity of rural poor to fend for themselves.

TENDENCY #4: Donor policies which foster dependence

Instead of empowering project members to achieve self-reliance, donor policies sometimes unintentionally foster dependence.

Villagers generally manage to survive despite their poverty and under-developed surroundings. Aid aims to improve their quality of life. The learning process findings indicate, however, that donor agencies sometimes unconsciously pursue policies or establish working rules which increase, rather than decrease, project members' continued dependence on outside aid. Project representatives at the regional workshops noted that donors sometimes provided too little funds for too short periods to enable the projects to get underway successfully. Donors often seek to end funding after a fixed period of, say, two or three years. That period may not be long enough, however, for project members to begin to produce a surplus for reinvestment. Projects may either collapse when the donor withdraws further aid, or apply to other donors for additional funds.

Other kinds of donor policies also tend to foster dependence. For example, in Zambia, a donor organization limited membership in a women's project to unemployed women. The women who originally organized the project, all of whom held jobs with the local government, therefore became members of the project's Board of Directors. The other members who joined the project, however, resented what they considered the denial of their right to run the project.

Another donor organization appointed a cooperant with 30 years experience in his home country as a steel worker, to work with the small-scale Zambian industry. His skill level exceeded that required by the project; the agency's rules, however, rendered the project members dependent on him in other respects. The agency prohibited anyone but the cooperant from driving the truck it had supplied to the project. Behind his back, the workers laughingly called him their truck driver. In fact, the industry depended entirely on him for transport of essential supplies and finished products. The cooperant did not speak the workers' language, so he could only talk to the manager who spoke English. It took him over a year to discover how the managerial committee mishandled grant funds (something which the student participant observer discovered in less than three weeks). Unable to communicate with the other workers, the cooperant then insisted on countersigning all the projects' checks drawn on the grants, creating another form of dependence. The unintended consequence of this was that, just as his metal-working skills were needed to train project members to use a sophisticated new piece of equipment provided by yet another donor, the managerial committee seemed likely to refuse to renew his contract.[10]

The Zimbabwean tractor scheme's experience suggests that donor organizations may assume that project members can deal effectively with new technologies before they have the necessary skills or access to essential complementary factors. Several members of the scheme expressed concern that the donor agency had not helped them to negotiate a sound contract with the US transnational corporation which, through an Indian affiliate, had sold them the tractors. The firm failed to set up a garage to service the tractors, although the members claimed that it had promised to do so. The project members had to transport broken tractors to the nearest repair station, located seventy miles away. As a result, broken tractors sometimes lay idle for months. The scheme's members argued the donor should have insisted the company carry out its promise. They also resented the high interest rates they had to pay the Agricultural Finance Corporation in accord with the agreement the donor agency had reached with that parastatal.

The Tanzanian intermediary agency that transferred donor resources in kind through the district officials to the Kigoma and Songea projects failed to build the members' capacity to plan and implement their projects.

Instead, the villagers remained dependent on the district authorities to acquire the appropriate inputs on time. When they failed to do so, the villagers blamed the district officials for their projects' failure. In Kigoma, the villagers claimed they had asked for assistance in building irrigation facilities to grow maize, not beans. They said they had never agreed to the use of oxen and would have preferred a tractor. The villagers maintained maize-bean cultivation conflicted both with their work on family plots and with the district's requirement that they also grow cotton. Lack of personnel and the location of the intermediary agency's head office in Dar es Salaam, 1000 miles away, hindered staff members from working directly with and getting to know the personal capability of the project members.

The working rules formulated by the donor agency for the food-for-development projects tended to leave the women recipients dependent on government and church leaders at the village level. Without consulting the women, the latter decided how to spend the projects' earnings.

A Tentative Explanation for Tendency #4

> *Donor organizations may formulate policies which on the surface appear justified, yet may foster project holders' dependence. Not infrequently, donor agency staff neither take the time nor possess the tools (knowledge of all aspects of the culture, language, socio-economic conditions) to involve project holders in a participatory process of assessing the obstacles and formulating more appropriate strategies.*[11]

In the Zambian uniform sewing case, the donor agency staff argued reasonably that it aimed to assist only marginalized women, so only those without jobs should work in the sewing project. Initially, apparently, they had not thought through the implications of the proposed rule leaving the women workers dependent on a non-working board of directors.

In the Zambian hoe-furniture case, the donor agency, again quite reasonably, sought to prolong the life of the truck it donated to the project by permitting only the volunteer cooperant to drive it. The cooperant, himself, concerned lest the managers pilfer donated funds, decided that he himself must countersign all checks. In both cases, however, the working rules fostered dependence and resentment among the workers and managers.

In the Zimbabwe tractor scheme, the project members apparently entertained unrealistic expectations of the donor. It may be also that the donor helped them to acquire capital-intensive machinery which they were not prepared to operate in a fully self-reliant manner. In any event, neither the donor staff nor the members appear to have adequately examined the objective factors likely to hinder the project's success.[12]

To explain why the Tanzanian intermediary agency concluded it had to rely on district authorities to dispense aid in kind requires more information about the agency itself. Unfortunately, the pilot Learning Process did not focus on the internal operations of the agency. On reviewing the findings, however, the final regional workshop did discuss the issues. The donor agencies that transfered resources through the intermediary agency benefitted from its reasonably close relations with the central and local governments. They could expect that projects receiving their aid would contribute to implementing national and district level development plans. The findings suggested, however, that the projects' objectives failed to coincide with the project members' perceived concerns and needs. By leaving the project members dependent on the district authorities, the agency neglected to ensure the members' participation in planning and implementation. The villagers at best worked only half-heartedly to achieve the projects' goals.

The second Tanzanian workshop pointed out that, where more than one institution has an interest in development in a single village, the situation may become more complex. As in Kigoma, each wishes to see "its" project succeed. The villagers, in contrast, prefer that whatever projects they undertake should contribute to overall village development. Thus the villagers suggested they be permitted to concentrate on growing either maize and beans, or cotton.

Suggested Strategy for Dealing with Tendency #4

> *To avoid unconscious policies that create project holder dependence, donor agencies should devise some form of participatory process through which they, together with the relevant intermediaries, work with project members (if necessary with the aid of local researchers), to scrutinize objective conditions and devise strategies which strengthen the members' capacity for achieving self-reliance.*

In several cases, the donor and intermediary agencies, working with the project members from the start, avoided creation of policies that fostered dependency. For example, in Zimbabwe, the leaders of the non-governmental umbrella organization and the women's community groups worked together with the donors to design appropriate resource transfers that strengthened, rather than inhibited, the members' self-reliance. The umbrella organization's staff planned with participating community groups in advance to design goals and working rules for the training institute. On this basis, they requested donors to finance the building materials with which the trainees constructed the institute building, and to hire qualified trainers. When donors offered funds for resources or personnel the villagers and the organizers considered inappropriate, they rejected the offers.

The women's groups elect representatives to administer the revolving credit fund. The women themselves assume responsibility for the fund. They determine the rules regarding repayment to ensure availability of funds for other projects. The women project members noted one problem, however: The rules they initially established restricted credit to new projects; this meant that on-going projects could not request more funds to expand. The women had begun to consider altering this rule. This illustrates the proposition that every new aid strategy—like proposals to resolve any human difficulty—will inevitably give rise to new problems.

In the integrated communal area program, the elected village committee's assistance to members of new income-generating projects in getting started with their own resources before requesting aid seemed to strengthen their own perception that they could be self-reliant. The leaders of over 90 percent of the projects indicated that, although they received aid from outside donors, they thought they could get on without it.

In others projects, the Learning Process helped donor and intermediary agency staff, as well as the project members, to work together more closely in adopting appropriate working rules. The Zambian women's intermediary worked with the women's uniform sewing group to formulate a constitution that outlined the respective responsibilities of the members and the board of directors. The donor agency staff helped the workers in the hoe-furniture industry to establish a workers' committee to meet and resolve issues of responsibility with the management. Several project members began to take bookkeeping courses.

The Zimbabwe tractor project members worked with the donor agency staff to establish a revolving fund to finance fuel, spare parts and tractor repairs. They arranged with the state farm for fuel and spares. They sent some cooperative members for training as mechanics. In addition, the members began to discuss with the donors how to change their working rules so that all the cooperative's members, including the poorer families, could use the tractors.

In the case of the food-for-development project, the donor agency began to involve the women themselves in implementing activities to increase production of their traditional foods, including improved water supplies, health and forestry. This helped the women reduce their dependency both on male village leaders and outside food aid.

The second Tanzanian workshop recommended that donor agencies should coordinate their aid plans to avoid unnecessary competition between projects for resources, and thus enhance overall village development. The workshop proposed that donor organizations might set up a coordinating office to foster institutionalized cooperation.

At the final regional workshop, a participant suggested the division of projects into those that generate incomes, and those that provide infrastructure, like the training institute. Where over time a project should become profitable, the donor might lend funds at a low rate of interest. Eventually, the project could repay the loan, making the funds available for additional projects. This seemed to work successfully with the women's revolving credit fund. For this strategy to work, however, the interest charged must remain low enough that it does not consume the project's potential investable surplus. Furthermore, the project design must ensure that members earn enough cash to live on. In the case of women's projects, women's marginalized status sometimes results in cash returns barely providing members with subsistence, far less a surplus. For projects like the Zambian small-scale industry or the Zimbabwe tractor scheme, on the other hand, as in Tanzania, government agencies could subsidize the provision of funds at rates of interest low enough to permit successful "takeoff." In these cases, donor agencies might, in the short run, provide a revolving credit fund. In the longer run, donor intermediaries and project members, together with national researchers, should press for establishment of government credit agencies designed to give low cost credit to viable small income-generating projects.

TENDENCY #5: Inadequate development policies

Chronic difficulties in obtaining essential inputs and marketing outputs.

The projects in all three countries confronted difficulties in finding adequate transport to obtain inputs and market outputs; getting and repairing necessary equipment and materials at prices they could afford; and storing, transporting and selling their produce at prices sufficient to provide them with the surpluses necessary to achieve self-reliance. Few of the income-generating projects could obtain adequate credit at low rates of interest. In Zimbabwe, projects in the Communal Areas (the former

"Tribal Trust Lands") faced additional structural obstacles caused by overcrowding and lack of access to fertile, well-watered land.

Innumerable difficulties of this kind disrupted all the projects' programs. For example, in the year between the time the Zambia hoe-furniture industry requested a donor to help them buy a welding machine, and the date when it finally arrived, its price had doubled, cutting the effective grant in half. The women's uniform-making project purchased seven sewing machines with their grant, but one broke during the rough trip to the remote western province. It took a month to get it back to the company for repair. The northern women's group had to rely on a woman who had her own transport to buy their materials in the provincial capitol, although they suspected her of keeping the difference between what she told them the material cost and the price she actually paid. The women's vegetable-growing project members complained that the farmer, who permitted them to use his irrigation pump if they purchased their own fuel, utilized the diesel fuel they had bought to pump water for his own crops.

In Zimbabwe, the high costs of imported fuel and parts, as well as difficulties of transport over rough dirt roads in remote communal areas, aggravated the other internal and external difficulties the projects faced. Rising prices slashed project members' real incomes. The members of the tractor project and the producer cooperative, both of which had borrowed funds to buy machinery and equipment, complained that they were "just working for the AFC" (Agricultural Finance Corporation); the 13% interest rate the AFC charged, they said, consumed all their profits.

Furthermore, the three-year drought hit harder in the communal lands where most rural poor still lived than in the better-watered commercial farm areas. For over a year, many project members survived primarily on drought rations.

In Tanzania, rural projects in remote, underdeveloped areas like Kigoma and Songea lacked transport to bring in consumer necessities and essential farm inputs, as well as to send to market whatever produce they succeeded in growing. Kigoma villagers complained that even if they grew surplus beans, maize or cotton as urged by district officials, they could not be sure the appropriate marketing boards would find transport to collect them. Furthermore, transport shortages hampered efforts of staff members of the intermediary agency, based in Dar es Salaam, to work more closely with project members. In the food-for-development project, instead of using the low cost food they acquired to feed their children, the women found it possible (and considered it necessary) to sell it in the open market to purchase high-priced necessities for all family members.

In short, in the 1980s, just keeping projects afloat became a major struggle.

A Tentative Explanation for Tendency #5

> *National development strategies have failed to sufficiently restructure inherited institutions and resource development patterns. These have left the independent southern African states externally dependent and hence especially vulnerable to international economic crises like that which gripped the world economy in the 1980s. Drought and South African destabilization policies, as well as conditions imposed by the International Monetary Fund, aggravated the impact of that crisis on the region's rural poor. All three countries have experienced rising prices and shortages of essential (typically imported or import-dependent) machinery, equipment, spare parts and fuel required by grassroots projects. This particularly aggravated the difficulties of rural projects in remote underdeveloped areas.*

Despite their differing ideologies, at independence all three newly-independent governments adopted the conventional wisdom that urged them to expand exports. If anything, these policies rendered their economies even more vulnerable to the international crisis that, in the late 1970s and early '80s, engulfed Africa. The terms of trade of all three countries sharply deteriorated, aggravating worsening balance of payments deficits. Lack of foreign exchange led to shortages of imported materials and equipment for import-dependent industries and agricultural projects, as well as spare parts and oil needed to keep the "modern" sector running. Revenues plummeted. The governments borrowed heavily, both internally and externally, to finance expanded social services and continued investments in economic infrastructure. (For background see Ch. 3 above.)

Zambia's and Zimbabwe's governments turned to the International Monetary Fund (IMF). As conditions for its aid, the IMF required them to implement policies that imposed the burden of the crisis on the lower income groups, including those in the rural areas whom most private voluntary organizations seek to assist. The IMF conditions included measures to:

i. reduce social service expenditures and state intervention to restructure the inherited political economy;

ii. hold the line on wages, cutting the workers' real incomes as prices rose;

iii. raise the interest rate, thus increasing the cost of borrowing for small businesses and peasants;

iv. expand trade with South Africa;[13] and

v. devalue national currencies, automatically raising costs of imports and spurring price increases throughout the economies.

Unemployment mounted. The crisis spread into the rural areas. The urban unemployed could no longer send home the earnings on which their families in the rural areas depended. Transport difficulties aggravated price increases and shortages, contributing to deteriorating living conditions in the countryside. In the early 1980s, prolonged drought, especially in the less developed provinces, required the import of foodstuffs, further accentuating the impact of the crisis. By 1983, about 40 percent of the inhabitants of Zimbabwe's still overcrowded Communal Areas—hardest hit by drought—survived only with the aid of relief provided by the government assisted by foreign donors. The rains finally came in 1984, but the economic crisis persisted.

South African tactics further destabilized both countries' efforts to develop. In 1979, the systematic bombing of major bridges was aimed to force Zambia, as a Front Line State, to accept the Lancaster House compromise constitution for Zimbabwe. This imposed heavy additional expenditures on Zambia's already hard-pressed economy. In Zimbabwe, South African undercover agents blew up a major ammunition dump, half the airforce, and the ruling party's headquarters. South African-trained and armed "contras" destroyed the oil pipeline and attacked the railroads carrying Zimbabwean goods through Mozambican ports; and killed people and attacked development activities in Matabeleland (see discussion of tendency #3 above).

The Tanzanian government also requested IMF help, but it rejected the IMF conditions. Therefore, the IMF refused to help Tanzania offset its balance of payments deficits. More serious, most other international public and private lending agencies followed the IMF's lead. Both the United States government and the World Bank slashed their aid to Tanzania. The Tanzanian government sought to implement its own structural adjustment program, and received some foreign assistance, especially from the Scandanavian countries. Nevertheless, Tanzania experienced exceptional difficulties in importing the fuel, spare parts and equipment for the transport and productive activities on which rural development projects depended.

The story of the impact of the crisis on vulnerable, externally dependent economies like those of Zambia, Zimbabwe and Tanzania is not new. The pilot southern African Learning Process, however, documented more details revealing the way the crisis aggravated the grassroots projects' problems. Shortages of machinery, spare parts and materials reduced output and employment in the so-called "modern" sector, adding to the growing numbers of un- or underemployed returning to the impoverished rural areas. The high cost of imported fuel and the lack of essential parts and equipment accentuated the problems of transport to previously neglected rural regions. For months, lack of transport facilities virtually cut off remote rural areas, hundreds of miles from the capitol, like Kigoma and Ruvuma. Combined with chronic shortages of foreign exchange, transport difficulties restricted the availability of all kinds of consumer goods, as well as the tools and equipment rural dwellers needed to improve their productivity and incomes. Those goods that finally arrived in rural shops cost two and sometimes three times the rising prices charged in the cities. The real living standards of the rural poor, which had risen significantly since independence, deteriorated.

In Zimbabwe, external pressures and contradictory tendencies within the new state aggravated these difficulties. International funds pledged to the Zimbabwe government to resettle peasant families from the Communal Lands to the underutilized commercial farm areas, failed to materialize. Yet even the previous government had estimated that twice as many people lived on these Communal Lands as they could support.

Furthermore, the inability of the government to implement its social reforms, including land redistribution, fostered widespread disillusionment. Taking advantage of this, South Africa helped to train and arm the dissidents in the western provinces. The government, claiming ZAPU supported the dissidents, declared a state of emergency and detained thousands of people.

SUGGESTED STRATEGY FOR DEALING

Suggested Strategy for Dealing with Tendency #5

This set of explanations suggests that small rural projects, alone, cannot devise strategies capable of avoiding the consequences of government policies that leave national political economies vulnerable to external pressures. In the long run, the countries themselves must devise plans and programs to reduce their nations' vulnerability to international crises. This underscores the need for Third World researchers and research institutions, together with local, national and regional planning authorities, to study the nature and consequences of alternative national and international development policies and their impact on rural peoples. By supporting institutionalization of the Learning Process, donor organizations may help them strengthen their capacity to formulate and implement improved national policies.

Most participants in the final regional workshop at Gwebi agreed that isolated projects can contribute little to overall improvements in community welfare. As an illustration, they discussed two Zimbabwe experiences introduced by a Zimbabwe intermediary participant. In one Zimbabweab village, the women learned to sew some uniforms and sell them. The women learned some new skills and earned a little cash. But this only marginally altered the lives of the poor majority. In a second village, the women, together with the other community members, planned the uniform sewing project as part of a larger community-wide program. All the villagers worked on various projects to raise the village income and improve everyones' living conditions. Some grew vegetables to contribute to better nutritional standards; others built food and water storage facilities to protect against drought; still others acquired skills to make and repair simple farm tools.

In conclusion, project members were urged to think through the relationship between their work and overall community development. Gwebi workshop participants pointed out, too, that increased members' understanding of their project's relationship, not only to their immediate village surroundings, but also to the larger world context, would empower them to design more effective development strategies.

The Gwebi workshop underscored the advantages of engaging national researchers in the pilot Learning Process. This, they pointed out, strengthened their capacity to understand the impact of national and international factors on grassroots projects. By supporting this linkage between researchers and peasants, the donor and intermediary agencies have taken an important first step to build up southern African capacity to plan more self-reliant national and regional, as well as local, policies.

At the same time, the workshop participants urged US private voluntary organizations to continue helping the citizens of their own countries to understand why the international crisis has had such a devastating impact on Third World countries like those in southern Africa. They should press the US government and international aid agencies to adopt policies which help, instead of hindering, countries like Zambia, Tanzania and Zimbabwe to restructure their inherited economies to meet the needs of their peoples, especially those in neglected rural areas.

TENDENCY #6: Special difficulties confronting women

Women's projects are particularly vulnerable to the consequences of the five tendencies discussed above.

In most of the projects in all three countries, the findings revealed systemic obstacles to rural women's participation in development projects. For example, in Zambia, the entrepreneurs who established the small-scale hoe-furniture industry employed only one woman as a clerical worker. They perceived the production work as a male occupation. The three relatively small Zambian women's projects seemed likely to contribute in only a limited way to surmounting the constraints that marginalized poor rural women. In all three cases, some members expressed resentment over the control over resources and decisions exercised by leaders who typically seemed to come from higher income, better-educated groups. All the women's groups seemed to fear domination by local government agencies. At the same time, some complained of lack of access to resources these agencies controlled. In the northeastern women's group, the women tended to shape their goals to meet their perceptions of the donors' requirements, rather than analyzing the factors thwarting their ability to resolve their own problems. In the crisis of the 1980s, all the women's groups faced aggravated difficulties in obtaining resources.

In planning their activities, four of the six Tanzanian projects neglected women almost entirely. From the outset, the carpentry and timber projects provided employment only for young men. In the two Kigoma maize-beans projects, the failure to involve women in initial planning undoubtedly contributed to less-than-satisfactory levels of productivity.[14] In the two food-for-development projects, although women were considered the primary beneficiaries, the project planners did not incorporate the women themselves in planning the project from the outset.

In Zimbabwe, four of the five projects claimed to address the issues affecting women as part of their larger community programs, arguing that women's status could best be improved in the context of overall community development. Women received equal voting rights, and participated among the top leadership in the umbrella community organization. Nevertheless, they did not appear to have achieved fully equal status in all respects. The leadership of the tractor cooperative included no women at all. The training institute trained women primarily for traditional roles—sewing and baking—while the men learned carpentry, construction, etc. Instead of expressing independent opinions, most of the women in the producer cooperative, mostly wives of male members, seemed to follow their husbands' lead in relation to controversial issues. In the integrated communal area program, men predominated among the leadership even in several projects in which women made up the majority of members.

The one Zimbabwe project designed primarily for women—the revolving credit fund—received considerable assistance from the women development officers, as well as sizeable contributions from a private voluntary agency.

The credit scheme could not, however, address the larger issue of the failure of land redistribution to resettle adequate numbers of peasant familes. This left rural women like these to wage a daily struggle to earn adequate incomes in marginal land areas.

A Tentative Explanation for Tendency #6

> *Attitudes and practices towards women, arising from traditional sexual divisions of labor, reinforced by colonial policies designed to coerce men into a low-cost laborforce, have typically excluded women from key decision-making positions and the mainstream of the development process. These attitudes and institutionalized practices have systematically denied women access to skills and resources required for so-called "modern" development, even at grassroots levels.*

This explanation coincides with the findings of the pilot process in relation to all the five tendencies discussed above.

1. The selection of projects

In Zambia, the women who chose to undertake sewing projects apparently adopted the traditional view of sewing as appropriate "women's work." In the western province, a market clearly existed for the uniforms they made. In the northeastern province, the women chose sewing despite the fact that they were not sure of its market potential. The church women who grow food to improve the nutrition of children in the rural Copper Belt again chose a project compatible with their traditional roles and existing skills. In the hoe-furniture industry, tradition apparently held that only men could or should do carpentry and metal work.[15] No one even considered the possibility of employing women in the productive aspects of the small industry.

In Tanzania, colonial rule treated Kigoma as a labor reserve for men who migrated to faraway sisal plantations, leaving women, children and old folk using outmoded hand tools to support themselves. In the 1980s, the district councils obtained financial assistance through the intermediary agency to create new job opportunities to attract and keep men in the area. In the south, the traditional attitudes of the villagers and intermediaries, if not the donor agencies, limited the employment by the timber and carpentry projects to young men. In Kigoma, the district council initially perceived the maize-bean project as part of integrated rural development designed to introduce oxen plows to improve productivity and raise community incomes, at least in part to keep the men at home. Though women cultivated much of the food crops, traditional attitudes excluded them from the village leadership. In neither of the two villages included in the Learning Process had the villagers elected a woman as one of the 10-house cell leaders who supposedly took part in selecting and designing the project.

In the Tanzania food-for-development project, the donor agency had consciously selected women as the "target population" to use a phrase common to the aid community.[16] Although the donor agency assigned a woman nutritionist from another region as the person responsible for the overall project, the village-level church leaders were all men. Some women did participate in the committees.

In Zimbabwe, the male leaders of the tractor project explained their exclusion of women from planning and decision-making by saying that other projects, like goat and poultry raising and uniform sewing—traditionally considered women's work—dealt with their concerns. It is hardly surprising, on the other hand, that the integrated communal area program appealed primarily to women who made up three-fourths of the projects' members. Because of the colonial institution of migrant labor, women constitute a majority of the inhabitants of most communal areas in Zimbabwe.

Not infrequently, women found that their work in projects added to the already double burden they carried: their care of children and the household, as well as their role in food farming. When the rains finally came in the credit scheme area, the women tended to neglect their cooperative gardens because they had to perform their traditional task of cultivating their family plots. In Tanzania, the women explained that they had difficulty finding time to work on the beans-maize projects, since they also had to take care of the family plots and the required cotton cultivation.

2. Formation of elites among project leaders

The women, who emerge as leaders in women's projects, usually have access to education and resources in part because of the income and status of male members of their families. In helping to organize the projects, they naturally draw on the resources available to them through their own connections to the "modern" sector, like church organizations, their husbands' activities, or their ability to communicate with donor groups. Their skills and access to these resources, however, may distance them from the poor rural women who make up the bulk of the project members.

Furthermore, the lack of bookkeeping skills limits the ability of other project members to monitor and ultimately participate more effectively as equals in running the projects. The traditional exclusion of rural women from educational institutions erects additional obstacles to their learning these essential skills. Since they cannot check the facts, they easily fall prey to suspicions that their leaders are manipulating the resources, including those provided by donor agencies, for their own benefit. Even if their suspicions are justified, they may fear to complain because they might lose the access to the resources their leaders enjoy.

In Zambia, the board of directors of the western province sewing group had education sufficient to obtain employment in the local administration. This precluded them from doing the actual sewing of uniforms. Until the project established working rules delineating their respective responsibilities, some members resented their apparent control. In the vegetable-raising group, the chairwoman's husband provided the land and the pump the women used to water it. Until they openly discussed and worked out a monitoring system, several members suspected he used the diesel fuel they purchased to irrigate his own land. In the northeastern women's project, the chairwoman's husband managed the coffee plantation on which many members or their husbands worked. This hindered the members from criticizing what some viewed as the chairwoman's misuse of the funds they had saved together. Unfortunately, unable to resolve this problem, the members drifted away from the project.

In several instances throughout the region, traditional factors seemed to have excluded women from leadership roles in the projects. In Tanzania, despite the official government policy of equality between the sexes, the 10 cell leaders in the rural areas included in the Learning Process were almost all men. Even in the food-for-development project, which primarily aimed at assisting women and their children, men controlled and directed the program.

In Zimbabwe, no women served on the leadership committee of the tractor project; the villagers assigned women to work on other, more traditional "women's tasks." In the integrated communal area program, in some projects where the majority of members were women, the leadership consisted primarily of men. The students reported that the projects did not seem to function as successfully as those where women participated more fully in the leadership. Although the students did not pursue the issue further, this finding suggests that, while various traditional factors may have led the women to vote for men as project leaders, they may not have responded as well to the men's leadership.

In the producer cooperative, only one woman—the wife of one of the original organizers—served for a time on the governing committee. In his discussions with the cooperative members, the student did not explore her role in depth. The student observed that both male and female members apparently adopted the culturally-accepted role of women; but unfortunately, he failed to explore the reasons more fully.

The women's revolving credit fund and the western Zimbabwe umbrella organization included women in the leadership. Although more research is required to determine the patterns of leadership-membership relations, the leaders in both cases clearly made special efforts to help the women members acquire literacy and other skills needed to monitor and participate effectively in decision-making at all levels.

3. Relations with district officials

The general exclusion of poor rural women from the development process typically creates situations where, in their attempts to gain access to resources, women confront largely male-dominated District Councils and other government agencies. This tends to generate among them greater fears of attempts by the local (largely male) elites to take over their projects. At the same time, it renders it more difficult for them to negotiate with government agencies for the kinds of resources the latter control.

These tendencies existed in Zambia, as suggested by the concerns of some of the women's projects noted above. Some members of two of the women's groups felt the district councils could have helped them more with housing, transport, and tractors.

In the Tanzanian projects included in the Learning Process, since women were generally excluded from leadership, women members typically did not participate directly in negotiations with the district authorities for resources.

In Zimbabwe, the tractor project, too, excluded women from the leadership, so they did not engage in negotiations with the state authorities. In the conflict with the Ministry of Lands and Rural Resettlement, the women in the producer cooperative generally adopted the same position as their husbands. On the other hand, the Ministry of Women and Community Development and other local officials gave special assistance to the women working with the revolving credit fund. The women leaders of the umbrella organization worked with the men to win district council support. In their case, however, the difficulties were compounded by the factors aggravating the ethnic conflict between the Ndebele and the Shona-dominated ruling political party.

4. Donor relationships

The fact that most poor rural women lack the skills needed to formulate effective proposals and communicate with potential donors, complicates the donor and intermediary agencies' difficulties in helping them design appropriate strategies. For example, in Zambia the female employees and wives of the coffee plantation workers tended to formulate their aims to coincide with those they thought donor organizations would prefer to fund. They failed to analyze the objective conditions that shaped realistic project possibilities. Instead, in hopes of receiving aid, they fabricated what the woman student who worked with them called "window dressing." In response to various visiting intermediaries' suggestions, they changed their aims several times, from working as a sewing group to earn income for themselves, to establishing pre-school facilities, to building a health center. The members of the women's vegetable project also had difficulty in clarifying their aims in terms of a realistic analysis of the basic constraints and resources with which they had to deal. To assist unemployed women establish an income-generating sewing project, a donor agency proposed that the already-employed women who started the project should constitute a board of directors. This initially caused resentment among some members.

In Tanzania, the donor and intermediary agencies' reliance on traditional leadership structures tended to leave women dependent on men who planned the projects. This remained true even when women were specifically the intended beneficiaries, or were expected to do much of the work. This undermined their effective participation. In Kigoma, they did not understand the maize-beans project; and, never having owned or handled oxen before, did not know how and were not expected to take care of the oxen. In the Dodoma food aid project, the women sometimes sold the food they acquired to obtain cash to buy consumer goods they considered more important; and suspected the leaders of misusing both the foodstuffs and the funds the project generated.

5. National development strategies

The historically-shaped exclusion of women from the development process in southern Africa aggravates their difficulties in combatting the consequences of national development policies which have left their countries vulnerable to negative international influences. In all three countries, colonial rule marginalized women and their families by systematically coercing male inhabitants of neglected hinterlands into low-paid labor reserves to produce exports. Post-independence policies focused on expanding crude exports, rendering their economies even more dependent on world markets dominated by transnational corporate conglomerates. Lacking education and the skills to obtain jobs in the so-called "modern sector," women remain locked into remote underdeveloped regions, hard hit by international recessions. Their low incomes render it impossible for them to buy tractors; and they typically lack the training required to operate and repair them if they could buy them. They seldom even own oxen. The crises of the 1980s aggravated all these factors.

In Zambia, shortage of fuel and spare parts and lack of transport caused added difficulties for all three rural women's groups. They had to rely on others for transport to buy materials, repair sewing machines, and sell their output. This tended to cause delays in getting materials or repairs. It fostered some members' suspicions that those who did their errands concealed the invoices of goods purchased, and the real income earned from those sold.

In Tanzania, the food-for-development project did not—could not—address the causes underlying the people's poverty in the central region: the failure of national policy, over 25 years, to end the nation's inherited, externally dependent economy. Instead of helping the new government implement strategies to reduce the country's external dependence, aid from the US, the World Bank and IMF emphasized rural development programs, like that introduced in the western province, designed to expand crude agricultural exports. When that strategy led to reduced national food supplies, US donor agencies became conduits for the disposal of surplus foodstuffs. Apparently, these were shipped to the country without adequate consideration of the consumption patterns or real needs of the women and their families among whom they were distributed.

In Zimbabwe, the women's revolving credit fund served to fund small projects of women living in an overcrowded infertile communal area. These conditions imposed serious limits on their efforts. The slowdown of the land resettlement program left the communal areas, where most rural women lived, still greatly overcrowded. Even interest-free credit could not help the women and their families escape these constraints.

Suggested Strategy for Overcoming Tendency #6

Donor and intermediary organizations have generally agreed to give special attention to assisting women overcome the historically-shaped traditions and institutions that exclude them from the development process. Nevertheless, both men and women project members need to engage in a systematic investigation of how, at all levels, these restrict rural women's opportunities. This knowledge would empower them to devise new ways to ensure that women, working as equal partners, help design and implement creative development projects.

Insofar as underdevelopment typically marginalizes women, they confront special difficulties in participating as equals in rural development. A Botswana study emphasized that, until the elimination of factors causing overall underdevelopment, women's projects will likely require continued outside assistance.[17] The Gwebi workshop participants emphasized that both women and men should view their projects as part of larger efforts to attain community development.

In Zambia, a concerned donor agency appointed a women's project officer who worked closely with a consultative group at the university and with the Learning Process to discover the particular nature of difficulties confronting women's projects. She worked with the Zambian researcher to select projects to help illuminate the special difficulties which plague women's groups in different stages of their development: prior to any transfer of resources (the northeast women's sewing project); after a year of aid (the rural Copper Belt vegetable-raising group); and after several years of assistance (the western province uniform sewing group). On the projects, the students worked with the members to negotiate with local officials, develop accounting skills, and formulate new working rules to resolve potential internal conflicts. The women's project officer drew on the Learning Process findings, not only to improve the strategies of the participating women's groups, but also to assist other women's groups. Her observations contributed, too, to deepening the insights relating to women and development obtained at the national and regional pilot process workshops.

The Zambian women's project officer made a special point of working with the women to overcome obstacles thwarting their access to local government resources. The male students who worked with two Zambian women's projects played a helpful intermediary role in negotiating improved relationships with local government agencies. In the case of the uniform sewing project, the student, together with the women's project officer, helped the women make more effective proposals for assistance from local agencies. In the vegetable growing project, the women invited the student to join them in approaching the neighboring army unit to obtain the use of a bulldozer to plow their fields.

In Tanzania, the donor agency had specifically designed the food-for-development program to help women and their children obtain low-cost foods to overcome serious malnutrition. Realizing that this was a new venture, the donor agency proposed to include the project in the Learning Process in order to discover what particular problems it confronted and how they might best be overcome. It appointed a representative to the Learning Process who made valuable contributions in helping to design both the first and the second national workshops. She worked with the district authorities to hold the second national workshop in the district, providing a unique opportunity for the participants to analyze the projects. She also helped to focus attention of the national and regional workshops on issues confronting women as well as the general problems of development. In electing her as chairperson of the weeklong final regional workshop, the participants from all three countries expressed their recognition of her contribution to the Learning Process.

In Zimbabwe, the woman national who served as project officer for the donor agency arranged to finance the revolving credit fund which specifically provided a means by which women could select and fund their own new activities. The introduction of training in bookkeeping skills, and the Ministry of Women and Development's employment of women community workers chosen by the women themselves, helped the women's projects to surmount the special constraints they faced. The women mainly chose to undertake traditional women's activities: vegetable-raising, baking, sewing. In a break with tradition, however, their evident success eventually attracted unemployed men, who asked to join them.

Likewise, leaders of the community umbrella organization worked closely with the members—both women and men—to ensure they had thought through all aspects of their group's program. They made a special effort to provide essential skills to all their members, including women, to enable them to participate in and monitor their own projects. They established the training institute for women as well as men, though women mainly learned traditional "women's skills." At the first pilot regional workshop, the training institute representative argued strongly that women should work together with men in community projects. He insisted that their relationship should be like that of a pair of pants and a belt: neither can get along without the other.

Although they disagreed whether women should organize separately or through community-wide programs, all the pilot Learning Process participants agreed that donor agencies should continue to work with female and male project members to explore the deeply-imbedded institutional and attitudinal obstacles blocking women's full participation in development activities. Only this knowledge, they concluded, would enable the members themselves to discover more fruitful avenues to giving women access to needed resources and the skills.

The final regional workshop recommended that donor agencies undertake investigations with national and regional institutions to explore ways of helping larger numbers of rural peasant women and their families to gain independent access to well-watered land and other resources to increase their productivity and incomes. Eventually, they might even resolve the debate which arose repeatedly in the Learning Process workshops as to whether aid to women is best provided through projects involving only women, or through community programs.

SUMMARY AND CONCLUSION

The evidence from the pilot Learning Process fieldwork shows that a participatory, problem-solving evaluation process can facilitate generation and validation of useful explanations and improved strategies for dealing with the systemic problems plaguing grassroots projects. Despite the different socio-economic contexts of the three countries, the pilot Learning Process found six interrelated clusters of systemic tendencies which, unless consciously avoided, may create serious problems:

1) Some aid projects exert only a peripheral impact on the poverty and underdevelopment engulfing the surrounding rural majority;

2) Because of their privileged access to resources, project leadership may acquire elite status and manipulate aid to their own advantage;

3) Local government officials may conflict with projects over the use of resources;

4) Some donors unconsciously and for seemingly good motives adopt policies which foster project members' continued dependence on aid;

5) National and international development policies may leave national economies vulnerable to external forces, causing chronic difficulties for small rural projects;

6) Historically-shaped attitudes and institutions erect special obstacles to women's efforts to participate in efforts to achieve self-reliant development.

Future participatory evaluations will undoubtedly disclose additional problems. Nevertheless, to the extent that the explanations generated by the Learning Process are valid, they expose constraints which project members must overcome, as well as resources on which they may draw, to devise increasingly self-reliant development strategies. In the last analysis, the tendencies identified reflect the inherent characteristics of underdevelopment which aid should help to overcome. The Learning Process findings suggest the advantages of involving the project members together with local researchers and intermediaries in designing, implementing, and monitoring their own integrated strategies for attaining development goals.

By enabling project members to work together with national researchers, including students from national research and teaching institutions, the pilot Learning Process showed that a participatory approach to evaluation makes two significant contributions:

First, it strengthens the members' own capacities to assess and handle new difficulties that inevitably arise as they work together to attain their goals. It also provides a structured environment in which they may learn from one another about useful strategies for solving the common problems they confront. In this way, it helps to empower the participants to achieve increasingly self-reliant development, one of the primary goals of aid.

Second, it adds to the body of knowledge available to national research and teaching institutions concerning the way development policies affect the lives and welfare of the poor rural majority. Hopefully, over time, this knowledge will empower national researchers to contribute to the planning and implementation of patterns of resource allocation and institutional changes capable of providing increasingly productive employment opportunities and rising living standards for all the peoples of southern Africa.

CHAPTER FIVE
Notes

[1]In the late 1970s, the World Bank undertook an integrated rural development program in Kigoma. This included encouraging the newly organized *ujamaa* villages to grow beans and maize, as well as cotton for export, along with building new roads, schools, and clinics. A Dar es Salaam University team, commissioned to evaluate the World Bank program, concluded that it had not achieved its goals of increased productivity, although it had contributed to providing needed social and economic infrastructure in the region. (See Anita Baltzersen, The World Bank and Problems of Rural Development. Worcester, Ma.: Clark University International Development and Social Change Program, unpublished Masters thesis, 1985).

[2]The difficulties experienced by the tractor scheme in Zimbabwe, however, would have been compounded by Tanzania's even greater foreign exchange constraints and the greater distance of the Kigoma villages from potential sources of fuel and repairs. Some local officials maintained the oxen had not received the necessary injections, whereas in the second village, where they had, they survived. However, the issue of the tsetse fly affects the entire southern African region, and requires regional attention at the government level for complete eradication.

[3]The donor agency staff initially had understood this small industry was a cooperative, but the Learning Process showed it was really a small private enterprise. With this new insight, the donor agency staff sought to help improve worker-management relationships.

[4]An eventual court trial proved false all allegations of misuse of funds.

[5]The ruling party of Tanzania.

[6]In the food-for-development case, the roles of the village and project leaders seemed less clearly defined.

[7]Established by the donor agency, following the first stage of the learning process, to fund fuel, spare parts and repairs.

[8]Later police investigations proved false their assertions that the leaders had threatened them with an AK47 and engaged in corrupt practices.

[9]Jonathan Steele "Pretoria's Secret War Against Zimbabwe," The Guardian (Manchester) Apr. 30, 1984.

[10]In the end, the cooperant had to leave due to illness so the matter never came to issue.

[11]The pilot Learning Process did not investigate the decision-making structure of the participating donor agencies that may have reflected typical centralized service delivery structures (See Chapter 2, p. 1).

[12]That the wealthier members used their profits from the scheme to buy cattle suggests the donor agency may have too quickly, without adequate research, accepted the tsetse fly argument to support purchase of tractors as opposed to less expensive and more easily managed animal-draft power.

[13]This was particularly true in the Zambia case; see Marcia M. Burdette, The Political Economy of Zambia (Lusaka: University of Zambia, mimeo, 1982).

[14]Juris Oldewelt, *Time Utilization of an African Peasantry: A Case Study from Kigoma, Tanzania* (Denmark: Centre for Development Research, 1984) showed that in Kigoma, in addition to their family chores, women do as much if not more farm work than men.

[15]Archaeological evidence, however, shows that in the pre-colonial division of labor, young women did much of the copper mining in the Zambia area.

[16]Women's groups throughout the world have noted that this concept typically contributes to patterns of behavior that exclude women from participation in planning projects affecting them.

[17]Women's Programme Development Project, *Women's Community Projects in Botswana, Recommendations and Strategies for Development Programmes*, (Gaborone, Botswana: Women's Programme Development Project, March 1985).

CHAPTER SIX

Toward Institutionalizing the Learning Process in Southern Africa

> The learning process makes a person know what she is doing. Otherwise, words just fly away.
>
> —A representative of a women's project at the Gwebi workshop, August, 1985.

At the final regional workshop at Gwebi, the participants concluded that the Learning Process could help to strengthen, not only the ability of project members, but also the capacity of the countries and the region to evaluate and improve the impact of aid. They proposed institutionalization of the process in the region.[1] At the same time, they recommended further participatory investigation to improve some features of the methodology, and noted issues needing further research.

This chapter summarizes the recommendations of the Gwebi workshop concerning the Learning Process. In keeping with the problem-solving approach, the chapter explains the causes of the difficulties encountered in the pilot process, and the suggestions made at Gwebi to surmount them. More specifically, the chapter:

- reviews the roles played by the various actors in the process (project representatives and members, national and student researchers, and the donor and intermediary agency staff) and suggests ways to improve their participation;
- considers ways of ensuring that the regional and national workshops provide an appropriate framework for introducing, designing, and evaluating the Learning Process methodology; and
- summarizes the conclusions of the Gwebi workshop as to how to improve the local, national and regional Learning Process.

THE ROLES OF THE VARIOUS ACTORS

From the beginning of the pilot Learning Process, the various sets of actors—project members, donor agencies and researchers—played different roles. The Gwebi workshop participants focused on ways to enhance the learning of each set and to improve their contributions to the process. They reached the following conclusions concerning the role of each of these actors, what they learned, and how their participation might be improved.

The project members

Their role in the process: First and foremost, the Learning Process aims to strengthen project members' capacity to criticize and improve their own efforts to attain self-reliant development. Therefore, the project members must play a primary role in implementing the Learning Process during all phases of the "project cycle:" (i) in making an initial feasibility study of their resources and possible obstacles to their success as the foundation for designing the project and formulating proposals for aid; (ii) in an on-going assessment of their project as a basis for continually revising and improving their strategies until the project reaches "maturity" (that is, until it no longer requires outside aid);[2] and (iii) thereafter as the project members continually strive to improve their own and their community's skills and resources.

The pilot Learning Process involved the members of grassroots projects in two ways. First, regional and national workshops brought representatives of the projects[3] together with donor agency staff and researchers to "learn-by-doing"—initially, by working together to design the participatory methodology; and, later, by sharing and comparing the findings, and evaluating the process. Second, project members worked together with university students (supervised by national researchers) to implement the process, analyze their own findings, and devise improved development strategies.

In the course of the year-long pilot Learning Process, these two procedures overlapped at many points. Each incorporated and built on the results of the other. The first national workshops served to introduce project representatives to the Learning Process and the other actors in it. The project representatives, in turn, introduced the students to the rest of the project members with whom they worked on the separate projects to implement the process. Together, the project representatives and the students evaluated their experiences in the second national workshops, and, in the Gwebi workshop, with project representatives from the other countries. Finally, in light of the debates and conclusions reached at these workshops, the members revised their own projects' programs and policies, sometimes with the further assistance of the donor agencies and the national researchers.

What they learned: The pilot process enabled project members to acquire two kinds of knowledge. First, they discovered better ways to obtain the technical skills and resources they needed to implement their projects. For example, the Zambian women, sewing uniforms, realized they needed to improve their ability to keep accounts. The Zambian researcher found an accounting student who spoke their language to live with them for several weeks to help them learn elementary bookkeeping skills. The Zimbabwe tractor project members realized that they needed to know how to repair their tractor, so the donor agency assisted them in sending two members for training. In several cases, as in the Tanzanian timber and Zambian vegetable raising projects, the members did not know how to sell their products. The students worked with them to investigate nearby markets and negotiate contracts with purchasers.

Second, the Learning Process awakened the project members to the fact that the problems they faced did not always arise from their own lack of technical skills or resources. The causes sometimes lay in the way they organized themselves, or even in the impact of poorly designed policies and programs adopted by national governments or international agencies. This knowledge stimulated them to discover new ways to work with each other and their communities. It sensitized them to the way increased stratification, or traditional attitudes excluding women, undermined their efforts. It encouraged them to join other groups to press for government policies and programs creating an environment more supportive of grassroots development.

For example, the Zambian hoe-furniture project, like other small industries, could not obtain credit from the banking system. The Gwebi regional workshop suggested that national research institutions might cooperate with the Small Industries Development Organization (SIDO) to undertake research concerning the possibility of establishing a government credit agency for small businesses. In Tanzania, the villagers learned that they all experienced troubles with district officials charged with distributing aid-in-kind. They agreed to support the intermediary agency's decision to decentralize so that the project officers could work more closely with the villagers in designing and implementing community programs.

The pilot Learning Process helped all the participants become more aware of the necessity of involving all project members, particularly women, at every stage of planning and carrying out their activities. The Gwebi workshop did not, however, resolve the debate raised by some project representatives as to whether women should organize into separate projects or participate fully and equally with men in community-wide projects. Some participants claimed that the findings showed male community project leaders not infrequently exclude women from the leadership and often from the benefits of projects, though they expect them to do much of the work; therefore, they insisted, women should organize their own income-generating projects to enhance their independence and improve their families' well-being. Others argued that, since women and men live in villages and work together in families, they have to work together to overcome the causes of poverty. Everyone at Gwebi agreed, however, that participation in the Learning Process could help project members realize that, to achieve their goals, they need to change attitudes and traditions that exclude women from decision-making positions; and stimulate them to find ways to enable women to share fully in all aspects of community projects.

National researchers

Their role: The southern Africans at the Gwebi workshop generally agreed that national researchers' facilitative role in the Learning Process should continue for at least three reasons.[4] First, many university researchers and students come from rural backgrounds and maintain their rural ties. Careful selection can easily find individuals who will willingly take part in and contribute to the Learning Process. Their knowledge of the specific circumstances of their countries, including the language and culture of rural communities, enables them to assist project members to comprehend and participate systematically in implementing a participatory problem-solving methodology. Furthermore, they may both bring to and learn from the Learning Process a broader understanding of the complicated interacting loca, national and international factors inhibiting rural development.

As a second reason for their participation in the Learning Process, the national researchers have important links with the national teaching and research institutions which should constitute vehicles of change in the larger society. On the one hand, knitting of closer ties with rural communities through the Learning Process the researchers will ensure that project members have access to the national and regional body of knowledge about the causes of the obstacles that block their progress. On the other hand, the researchers will learn the practical hands-on kind of information those institutions need to improve their contribution to national policies affecting rural development.

To illustrate: The pilot Learning Process findings suggest the need for all kinds of further research. To note only three, the findings underscore the need to discover how to reduce the costs of credit for small peasants and producer cooperatives; involve women more effectively in decision-making roles in village institutions; and help small rural industries contribute to more balanced, integrated national and regional economies capable of providing more employment opportunities and raising living standards. Hopefully, findings like these will stimulate national research and teaching institutions to investigate and propose government as well as non-government measures to help solve very real rural problems which communities in underdeveloped rural areas, alone, cannot handle.

As the third reason for their participation in the Learning Process, national researchers may bring the findings to test and deepen the theories that guide national planning and the training of future planners. In all three countries, the Learning Process generated a body of new information about rural projects which university lecturers could give to students to help them understand the obstacles to rural development. Over time, incorporation of these kinds of findings may help to improve national planning to provide environments more supportive of rural folks' efforts to attain sustained self-reliant development.

Although emphasizing the value of links between individual researchers and national institutions, the Gwebi workshop participants cautioned against tying up the process in bureaucratic red tape. Future Learning Process organizers will need to seek out researchers to work as individuals with the Learning Process, while encouraging them to retain and strengthen their relationship to their institutional base. How this works in practice might vary from country to country and institution to institution.

Improving their role: The Gwebi workshop considered several aspects of the researchers' roles as facilitators in the participatory process. The participants discussed how the researchers could help project members understand and develop the learning-by-doing process. For more than half of their lives, educational institutions have socialized people to the notion that teachers stand at the apex of a hierarchical pyramid. In the Learning Process, in contrast, researchers need work in a "horizontal" learning relationship, as equals with villagers. The African Participatory Research Network, which has for several years been exploring ways of improving participatory research, might help improve the theoretical role and practical work of researchers.[5]

The selection and training of the students

Selection: The national researchers must select and train students who will live and work with the project members. Since 70-80 percent of all southern African university students come from rural communities, the national researchers found many individuals with relevant backgrounds and skills who empathized with villagers. Special difficulties arose, however, in finding women students with the relevant cultural and language backgrounds. In the less developed rural areas where many of the projects operate, far fewer southern African women than men enter universities. The Tanzanian researcher had to ask a woman student from another part of the country to work with the one qualified female student he could find from Kigoma. For two of the three Zambian women's projects, the national researcher, in consultation with the women's project officer, had to chose male students with the relevant backgrounds.

Although the first regional workshop in Lusaka had concluded that women students would find it easier to work more closely with women project members, the use of men did not turn out too badly in the pilot process. Women students tended to find it difficult to socialize with men in order to help them get to the roots of the problems plaguing the projects. Since men's traditional attitudes posed obstacles to women's full participation in community activities, male students were able to help male project members gain greater insight into why they should change them. In the Zimbabwe projects, however, though the male students noted the exclusion of women, they did not pursue the matter to discover the reasons. This suggests that the issue is not merely one of selecting appropriate students—male or female—but also one of focusing the students' attention more on the obstacles to rural women's participation.

Training: The Gwebi workshop participants noted that the national researchers should train the students to work more systematically with the project members to gather evidence. The students need to learn to work with project members to gather and analyze more fully the background data relating to the nature and needs of the community, the origins of the project, and a profile of the project members and their families. One project representative put it bluntly: "They should get the full story from all the project members, not just fibs from a few individuals who may be sell-outs."

The students should read and discuss the available material relating to the problem areas where the pilot Learning Process suggests systemic tendencies will likely plague all projects. This may help them to work together with the project members in discovering causes for similar difficulties, and devising better ways to deal with them. Before they go to the field, the students should have met at the national workshops with the project representatives to discuss the contexts and the problems that concern the members. Together with the project representatives, intermediary staff and national researchers, they should agree on the kinds of information needed to test the range of possible explanations. This may help ensure uniformity of standards of evaluation. Once they begin working on the project, however, the students should remain sensitive to the members' desires to explore some issues more fully than others to discover the underlying causes of the problems as they view them. This is especially true if (as the pilot Learning Process suggests is often the case) the members perceive the difficulties differently from the project representatives. This may prove particularly illuminating as to the underlying causes of the problems plaguing a particular project.

As an initial guide, however, the pilot Learning Process findings suggest that the students' training should sensitize them to the following kinds of questions (the regional and national workshops should review and revise this list in light of their particular concerns):

1. THE PROJECT'S ROLE IN THE COMMUNITY: Do the stated goals respond to the members' needs? To the needs of the broader community? Who decided on the goals? Do all the members agree with the goals? What is the sex, class, ethnic and religious background of the members of the project compared to those of the surrounding community? Do they all work effectively together to achieve the goals? Do they have the needed resources?

2. INTERNAL PROJECT RELATIONSHIPS: How are decisions made in the project? How do the leaders relate to the members? Who participates in the project's activities? What skills do members need in order to participate more effectively in implementing and monitoring the project's progress?

3. RELATIONS WITH GOVERNMENT AUTHORITIES: How does the project relate to the larger community and possible patterns of stratification within that community? Do local authorities have resources that might benefit the project? How are decisions made relating to these resources? May project members achieve improved access to those resources without disadvantaging other community members? How?

4. RELATIONS TO DONOR AGENCIES: How do donor agencies decide what aid to provide? What role do project members play in this process? Does the aid meet the projects' needs? Do the working rules accompanying the aid foster project members' increased self-reliance?

5. THE IMPACT OF NATIONAL AND INTERNATIONAL POLICIES: How do national and international policies and programs impact on project success or failure? How are decisions made concerning these policies? How and to what extent can project members influence them? Can the project members revise their development strategy to reduce their dependence on national and international factors?

6. THE PROJECT'S EFFECT ON RURAL WOMEN: Are women expected to work on the project? Will they benefit from it? Do they participate in the decision-making processes? What steps could project

members take to improve women's participation in all aspects, including decision-making and the benefits? What is the project's impact on women's income? Status? Relationship to the community? What effects do national and international policies have for the project?

The Gwebi workshop recommended that national research and teaching institutions direct greater attention to training students to ensure that they participate effectively with project members to realize the aims of the Learning Process. Wherever possible, the national researchers should help the students receive course credit for their participation in the Learning Process. To help prepare students, the university staff might develop appropriate courses in participatory, problem-solving evaluation methodologies. Lecturers might assign students to write background papers on the history and socio-economic features of the rural areas in which they work. Where senior students do honors theses, as in the University of Zambia, they could analyze the project as a case study. University staff should create opportunities for the students to share their findings with their classmates as valuable evidence for testing the theories of rural development found in the literature.

The donor and intermediary agencies

What they gain from the process: Donor agencies and their intermediaries stand to gain from an on-going participatory problem-solving evaluation of projects they assist. Insofar as it strengthens the project members' capacity to conduct self-evaluation to increase their self-reliance, it contributes to attainment of most donor agencies' stated aim. It should provide donor and intermediary agencies with a "window" on the functioning of projects so they can work with project members to devise more effective aid programs. In the process, their staff members should learn from national and regional research-teaching institutions about the likely consequences of alternative kinds of aid in particular socio-economic backgrounds. On the other hand, those institutions may learn from donor agencies' knowledge about grassroots development, drawn from their experience in the region and elsewhere.

Some further suggestions: The donor and intermediary agency representatives in the pilot Learning Project noted that non-governmental agencies often work in a relatively uncoordinated way in the same field, and even with the same projects. Each agency introduces its own policies and requirements. This occasionally introduces an element of competition between them. Their sometimes conflicting approaches aggravate the project members' difficulties in maximizing the benefits from their aid. The intermediary agencies' personnel reported the pilot Learning Process helped them to coordinate their efforts to work with projects.

As noted in Chapter Two, the pilot Learning Process did not explore the way the donor agencies' internal structures influence their aid decisions. The Gwebi workshop suggested that this might prove a fruitful exercise. Involving project representatives might also enhance their understanding of the reasons for particular aid policies. This would enable them to assess more realistically the implications of those decisions for their own activities.

The Gwebi workshop urged the donor and intermediary agencies to help institutionalize the proposed on-going Learning Process in southern Africa. Some have already set aside a small percentage of their budgets for research and evaluation. The Gwebi workshop participants voiced the sincere hope that in southern Africa they would contribute a significant portion of these funds to help finance the regional process. This would help to employ part-time national and regional coordinators, and bring representatives of projects, donor agencies, and research and teaching institutions together in workshops to work out the organizational details. Once the process is institutionalized, the donor agencies could help the projects they assist to participate in it on a fee-for-service basis. By providing this kind of support instead of continuing to employ outside "experts," the international aid community would make a valuable contribution to building the national and regional evaluation capacity essential to achieving self-reliant development.

In sum, each set of actors learned enough in the course of the pilot Learning Process to make it worth institutionalizing it in the region. In so doing, however, each will need to improve the way they contribute to the overall process.

THE WORKSHOPS

The Gwebi workshop underscored that during the pilot Learning Process the national and regional workshops created a valuable learning environment. Nevertheless, the conduct of the workshops left unanswered several procedural and methodological questions. These included: 1) Should one or two representatives from each project participate in the workshops? 2) Why not include the students in the regional workshops? 3) Would it be better only to include representatives from similar kinds of projects in workshops to design the details of the participatory evaluation process? 4) What is the best way to help workshop participants to understand and learn to use a problem-solving methodology? 5) How should workshop participants explore the appropriate analytical framework locating rural projects in the world system? 6) Should the Learning Process coordinators circulate written reports before each session?

Everyone who took part in the workshops learned a lot from the others. At one level, the project representatives shared new ideas about how to improve their projects' activities. They even learned new ways of making items for sale or for home use. The Zambian project representatives brought samples of their work—hoes, uniforms, and carry-all bags—and exchanged ideas with the Zimbabweans and Tanzanians about how to make these kinds of items. For example, when the Gwebi workshop participants visited the Zimbabwean tractor project, they learned from the villagers how to dry slices of tomatoes in the sun to preserve them for use in dry-season cooking. Back in Gwebi the next day, several people suggested that more regional workshops, accompanied by visits to local projects, might spur greater exchange of production techniques and even goods between grassroots projects, not only within each country, but throughout the southern African region.

More important, the workshops themselves constituted a participatory framework for learning-by-doing. The project representatives from Zimbabwe, who did not have the opportunity to take part in either an introductory or a final national workshop, expressed their opinion that they had missed an important feature of the pilot project. The first set of regional and national workshops gave the project representatives, the relevant intermediary staff members, and the national researchers an opportunity to work together to understand and adapt a participatory problem-solving approach to evaluation. The second set enabled them to share, compare and analyze the results of their year-long experience. In both sets, they obtained new insights as to the causes of the difficulties they faced, and learned from one another about possible new methods for coping with them.

The Gwebi workshop noted several matters, however, that require further consideration and experimentation. First, who should attend the workshops? Certainly they should include representatives of all the projects, relevant donor agencies, and national researchers, who participate in the particular Learning Process. But should two representatives, rather than one as in the pilot process, come from each project? This would make the workshops larger and more unwieldly, as well as more expensive. It would, however, bring to the workshop more than one picture of each project; and, on their return, two reporters, rather than one, might give the project members a fuller understanding of the workshop proceedings.

Most people at Gwebi thought that the students should attend not only the national, but also the regional workshops. This would extend the dialogue between them and the project representatives, facilitating the on-going sharing and comparison of the findings on a regional level. The students could both contribute to and learn from regional workshop debates.

The students might also help deal with the problem of translation which, in various ways, plagued both of the regional workshops. From the outset of the pilot Learning Process, a contradiction appeared evident: The pilot process aimed to involve representatives of poor rural southern African projects. But many of them could not speak English. How could they take part in the regional workshops?[6] At Gwebi, a village teacher represented the Kigoma projects in Tanzania because neither of the village representatives—including the one who had chaired the Tanzanian national workshop, conducted in Swahili—could speak English. The Zambian women's project officer successfully translated for the representative of one of the women projects, but that responsibility inhibited her own participation. By serving as translators, the students could facilitate project representatives' participation.

Some people recommended the organization of the workshops around the concerns of similar kinds of projects: credit, women's income-generating activities, producer cooperatives, etc. But questions arose as to how to arrange this. Would a threshhold number of similar groups first have to recognize the benefits of engaging in a participatory evaluation? Or might the more typical workshops, built around issues, incorporate the design of an on-going Learning Process in their activities, to be followed a year later by a second workshop to review the results? At Gwebi, most project representatives asserted that the second national and regional workshops, following their intensive year-long evaluation process, did much more to help them solve the problems hindering their work than the usual one-time-only gatherings.

Everyone at Gwebi agreed that the regional workshops should be extended to include representatives from other SADCC countries. At Gwebi, people regretted the absence of the researcher from Botswana and project representatives from ANC and SWAPO[7] who had attended the Lusaka workshop. Inclusion of participants from Portuguese-speaking countries would further complicate the issue of language; nevertheless, the Gwebi workshop concluded that the future Learning Process should seek to include projects and participants from all the southern African countries. Initially, as part of institutionalizing the Learning Process, some might be invited as observers to the regional workshops. As quickly as possible, however, the process should extend to include projects and participants from all of southern Africa.

The Gwebi workshop also explored several methodological issues related to the organization of the workshops. The Lusaka workshop debated how to introduce the problem-solving methodology. Most people rejected the use of lectures that project representatives might find abstract, incomprehensible and boring. Someone skilled in participatory research might lead the workshop in a two-way dialogue to explore how various aspects of the method might be used to deal with the projects' problems. The participants could actively engage in comparing it to the way they ordinarily assess their own work. They could use role-playing to illustrate how they might use a problem-solving on their own projects. Perhaps a film could be prepared to show how, in different rural settings, project members worked together with students to investigate and solve real problems. Perhaps a combination of these methods might serve. Future Learning Process participants will need to experiment with various ways of learning how to adapt a participatory, problem-solving approach.

A related methodological question was—how should the workshop participants explore and agree on the appropriate analytical framework to illuminate the way their projects and the surrounding underdeveloped communities relate to the world system? At what point and how should national researchers seek to introduce models (like that in Figure 3.1) to help project members understand why and how national and international policies influence their chances for success? When and how should national researchers submit their broader perceptions of these relationships for project representatives to critique in light of their grassroots experience? The regional workshops left these questions for future Learning Process participants to tackle.

The Gwebi workshop briefly debated a few additional matters. Some people suggested that, to reduce the time devoted to verbal presentations, written national reports should have circulated before the regional workshops. Future Learning Process participants will need to determine whether the constraints of time, financial resources, literacy and language barriers make this possible or desirable.

A future Learning Process will need to think through all these and a host of similar issues.

SUMMARY

The Gwebi workshop recommended institutionalization of the Learning Process to evaluate aid's impact on southern African rural development. Whether they came from remote rural projects, intermediary agencies, or national universities, the workshop participants concluded that the pilot

Box 6.1: Summary of the Gwebi Workshop Recommendations to Improve the Learning Process

The Gwebi workshop participants proposed the institutionalization of a participatory, problem-solving Learning Process in southern Africa. This, they concluded, would strengthen the capacity of rural project members, working together with national research and teaching institutions—as well as donor agencies—to ensure that aid contributes to increased self-reliant development.

At the same time, the Gwebi workshop participants suggested ways of improving the role of the four sets of actors and the function of the Learning Process during all phases of the project cycle to:

Project members
To strengthen their capacity to criticize and improve their own efforts to attain self-reliance, project members should participate in the Learning Process during all phases of their project cycle to:

- undertake the initial feasibility study of their resources and the possible obstacles confronting them as the basis for designing their project;

- evaluate their implementation of their project to revise and improve their strategies until they no longer require outside aid; and

- continue to engage in the Learning Process even after their project becomes fully self-reliant, working with each other and other members of the local, national, and regional community to critically evaluate and improve their development efforts.

National researchers
To help the project members strengthen the Learning Process, as well as to learn from it, national researchers should play a role as facilitators to:

- think through and help to resolve the theoretical and practical issues involved in developing and implementing the Learning Process methodology;

- inform and help Learning Process participants deal with the impact of local, national and international factors on development at the grassroots;

(Continued)

Box 6.1: *Continued.*

- incorporate the on-going Learning Process findings into the body of knowledge disseminated by national teaching and research institutions to improve the theory and practice of national planning to build an environment supportive of rural development; and

- select qualified student researchers and provide an appropriate training program for them.

Students

Living and working directly with the project members for prolonged periods, the students constitute a key link between them and the national teaching and research institutions in the development and implementation of the Learning Process. Their selection and training should ensure that they:

- speak the language and understand the culture of the project members;

- fully understand the participatory problem-solving methodology as a crucial ingredient of an effective Learning Process;

- work closely with, and learn alongside, the project members about the nature and causes of the difficulties plaguing their projects as a foundation for designing better project strategies;

- comprehend and introduce for project members' consideration the implications of national and international factors that may influence their projects to enable the members to devise strategies to deal with them; and

- become sensitive to and assist project members to investigate the consequences of exclusion of women in blocking rural development so they will search for ways to ensure women's full participation in all aspects of their projects.

Private voluntary organizations and intermediary agencies

As the source of the potentially useful transfer of resources that constitute "aid," private voluntary organizations and the intermediary agencies through which they operate may play an important role in building the regional Learning Process. To do so, they should:

- learn how to improve the impact of the resources they provide by facilitating project members' full participation through the Learning

Process in examining all aspects of the design, implementation, and evaluation of their projects;

- work with national researchers (rather than foreign "experts") to learn with and from them about the interaction of local, national and international factors that shape the impact of aid. This should deepen national theoretical insights and practical measures affecting the possibilities of self-reliant rural development; and

- consider ways of enabling grassroots project members to participate in evaluating their own procedures, as donor agencies, in making decisions concerning the transfer of aid.

The national and regional workshops

The national and regional workshops create a framework for enabling the participants to learn-by-doing how to design and implement the Learning Process to meet their needs. Through the workshops, the participants may learn from each other by sharing and comparing their experiences. Their formal and informal interactions should empower them to build networks to strengthen not only their project level activities, but also their national and regional programs to attain greater self-reliance. Over time, to achieve this, the Learning Process should extend to include all the independent southern African states. At the same time, future organizers of the regional Learning Process should devote attention to:

- organizing future workshops, and exploring the possibility of introducing the Learning Process into the numerous workshops that already take place. This would enable the workshop participants to engage in the systematic, on-going evaluation, improvement, and exchange of ideas concerning their efforts to deal with the issues with which these kinds of workshops grapple;

- introducing the Learning Process to enable participants to understand the methodology and the underlying analytical framework, not through dry, abstract lectures, but an exciting, learning-by-doing process; and

- improving the workshops to ensure the national researchers and students, as well as donor agency staff members, learn to work together with project members at every level to improve and strengthen the national and regional Learning Process.

Learning Process had deepened their insights into the problems they confronted, and facilitated efforts to devise more self-reliant development programs. The workshop emphasized the benefits of sharing and comparing their year-long learning experiences through national and regional workshops.

The Gwebi workshop stressed, however, the importance of further participatory investigation to improve the role of various actors in the Learning Process. Participants focused on the need to improve training for the students, and to strengthen the links between the national teaching-research community and members of grassroots projects. They suggested further consideration of how the workshops might more effectively create an opportunity for all participants to learn-through-doing to adapt a problem-solving methodology to meet their own needs. In resolving these and other difficulties encountered in implementing the pilot process, the Gwebi workshop participants concluded that a future regional Learning Process could contribute much to strengthening the national and regional capacity to achieve self-reliant development.

CONCLUSION

After more than two decades of independence, despite growing amounts of aid, widespread crises prevailed among most if not all of the new nations of Africa. Critics suggested that donor agencies failed adequately to take into consideration the needs and concerns of those to whom they provided assistance, and that the conditions surrounding aid sometimes undermined project members' capacity for self-reliant development.

The Southern African Pilot Learning Process aimed to empower the participating project members, working together and with national researchers and intermediary agency representatives, to design more effective strategies to overcome the obstacles to their own efforts to improve the quality of their lives. At the same time, it sought to provide the project members, intermediaries and national researchers, and the donors with a "window" into the much larger issues of the way aid impacts on grassroots development.

At Gwebi, however, the workshop participants underscored the necessity of viewing the Learning Process not as a single event with a beginning and an end, but as an on-going activity. They voiced their confidence that, in building on-going Learning Process activity, project members, together with national researchers and private voluntary organizations, could achieve more effective rural development. Working through a regional network, they could learn more about ways to strengthen not only grassroots projects, but also national and even regional self-reliance.

CHAPTER SIX
Notes

[1]Institutionalization here means creating a network of researchers and intermediaries prepared to work with project members to design and implement participatory evaluations of aid and development.

[2]The timing of the project cycle varies from project to project, influencing the length of time the project holders may require outside assistance. Given the marginalization of the poor, especially women, some kinds of projects— like literacy programs or child care— may never achieve "independence" of aid except in the sense that a (new or changed?) state institution may assume responsibility for them. (See Tardep's Women Development Programme, Evaluation of Activities During the first Five Years— Options for Further Planning, Report of talks held between the team of Tardep's Women Development Program and Johane Cottenie at Tarime, Tanzania, Aug 25-27, 1983).

[3]See chapter four for method of selection of projects from which the representatives came to the workshops.

[4]Some argued that, in countries like Dominica and India, university staff and students generally did not sufficiently empathize with grassroots projects' difficulties; and therefore others would have to play the facilitative role.

[5]At its 1985 Arusha workshop, the African Participatory Research Network began to address the issue of how to use this method to evaluate aid. Its executive secretary, Kamal Mustafa, a Sociology lecturer in the University of Dar es Salaam, Tanzania, reported that the network would pursue this task in the context of the proposed institutionalization of the Learning Process.

[6]Simultaneous translation facilities were prohibitively expensive.

[7]The African National Congress (of South Africa) and the South West African Peoples' Organization (of Namibia).

Participants discuss a knotty problem.

Index

Appendix: The Pilot Learning Process Advisory Board

Advisory Committee to the Development Strategy Assessment Project:

Lawrence R. Simon, Chair

Elizabeth Coit
: Associate Director for International Programs, Unitarian Universalist Service Committee

Richard Ford
: Co-Director of International Development and Social Change Program, Clark University

Catherine Gwin
: Independent Consultant, formerly Senior Associate, Carnegie Endowment for International Peace

Albert Hirshman
: Professor of Social Sciences, Institute for Advanced Study, Princeton University

Bernard Magubane
: Professor of Anthropology, University of Connecticut

Robert O'Brien
: Executive Director, Private Agencies Collaborating Together

John Sommer
: Dean of Academic Studies Abroad, Experiment in International Living

Richard H. Ullman
: Professor of International Affairs, Woodrow Wilson School, Princeton University

Antonia Hernandez
: Vice-President for Public Relations, Mexican American Legal Defense and Educational Fund

Brian Smith
: Professor of Political Science, Massachusetts Institute of Technology

Corresponding Member

Paulo Greire
Director of the Centro de Estudos em Educacoa in Brazil